COUNTDOWN TO GCSE

BIOLOGY

Sallie Powell

Head of Biology, Harrogate Grammar School

M

MACMILLAN
EDUCATION

First published 1987

Published by
MACMILLAN EDUCATION LTD
Houndmills, Basingstoke, Hampshire RG21 2XS
and London
Companies and representatives
throughout the world

Printed in Great Britain by
Cox & Wyman Ltd
Reading

Designed by
Plum Design
Southampton

British Library Cataloguing in Publication Data
Powell, Sallie
Countdown to GCSE: Biology
1. Biology
I. Title
574 QH308.7
ISBN 0-333-41858-1

ACKNOWLEDGEMENTS

The author and publishers wish to acknowledge the following sources:

HMSO for extracts from The National Criteria by permission of the Controller of Her Majesty's Stationery Office; also London and East Anglian Examining Group; Midland Examining Group; Northern Examining Association; Southern Examining Group; Welsh Joint Education Committee.

The publishers have made every effort to trace the copyright holders, but if they have inadvertently overlooked any, they will be pleased to make the necessary arrangements at the first opportunity.

CONTENTS

Countdown to GCSE: Biology

Introduction 1

1 Aims and objectives of the new exam 3

2 Syllabus content

LEAG draft syllabus 7
MEG approved syllabus 9
NEA approved syllabus 12
SEG approved syllabus 14
WJEC approved syllabus 16

3 Preparing for the exam

Your study schedule 18
Organising your revision 24

4 Sample questions

NEA specimen papers 29
SEG specimen papers 41
LEAG specimen papers 50
MEG specimen papers 53
WJEC specimen paper GCE/CSE 57
Index to questions 61

5 Specimen answers

SEG specimen paper 63
LEAG specimen papers 67
WJEC specimen papers GCE/CSE 74

6 Assessment and marking

Assessment 77
Grade descriptions 80

Summary 82

INTRODUCTION

If you are being entered for a new GCSE course, you are probably asking, 'Why have they suddenly started a new exam?' or 'What was wrong with the GCE anyway?'

To answer the second question first, one of the problems with the GCE and CSE exams was that they were run by twenty-two different exam boards, each with one or more different syllabuses. Although each was called 'Biology', the content of some of the syllabuses was very different.

The standards of the different boards also varied. If a school had entered the same candidates for exams set by different boards, it might have found that they were awarded different grades. If you told a prospective employer that you had a grade C in Biology, there was no absolute measure of what you knew or what standard you had reached.

The fact that there were two separate sets of exams meant that a teacher had to decide which course was most suitable for a pupil two or three years before he or she was due to sit the exam. It is not easy to predict how a thirteen-year-old is going to develop; too often mistakes were made.

The decision to replace the dual system of examining with a single examination has not been a sudden one. Trial examinations were first introduced up and down the country more than fifteen years ago. The experience gained from these 16+ or GCE/CSE exams has been used in setting up the GCSE.

In February 1980, the Secretary of State for Education and Science invited the examining boards to send representatives to form a Joint Council to replace the GCE 'O' level and CSE examinations with a common system of examining: the GCSE. The role of the Joint Council was to prepare draft *national criteria* to act as a basis for the GCSE exams.

The twenty-two examining boards were divided into five groups, each submitting its own syllabuses and setting its own exams. These five groups are:

London and East Anglian Examining Group (LEAG);
Midland Examining Group (MEG);
Northern Examining Association (NEA);
Southern Examining Group (SEG);
Welsh Joint Education Committee (WJEC).

These groups will be referred to by their initials throughout the text.

The Joint Council set up working parties for each subject. They listed the aims and objectives for GCSE Biology which are common to all the examining groups. They decided on a minimum common content of the syllabus. At least two-thirds of each syllabus must be made up of this core; the remaining third allows some room for innovation and experimentation by the groups. But the application of biology to technology and society is emphasised throughout the syllabuses.

It is the approach to the content which is different in the new exam. This is a practical approach, in which the skills and scientific processes learnt in studying biology carry as much weight in the final assessment as the student's knowledge and understanding of the subject. So the grade you will be awarded depends on the extent to which you have met the objectives of the exam.

This is a different method of grading to that of the GCE 'O' level, where the candidates were arranged in order of merit and the board decided on a cut-off point for each grade.

In the GCSE, if you have achieved the minimum standard for a particular grade, you will be awarded that grade irrespective of the performance of other candidates.

So the new exam does seem to be an improvement on the previous system, in that GCSE is a single system of examination fulfilling the demands of the National Criteria. Let's have a look at it in more detail, beginning with the aims and objectives.

SECTION 1

Aims and objectives of the new exam

Each of the GCSE syllabuses with the title 'Biology' embodies the aims and objectives of the National Criteria for Biology. If you examine these aims, you can see what you should expect to gain from your course of study.

Let's look at what the aims mean in practice.

	Aims	*Practical experience*
1	**To develop an interest in, and enjoyment of, the study of living things.**	To study a variety of plants and animals in order to stimulate a life-long interest in all living things.
2	**To encourage an attitude of curiosity and scientific enquiry.**	To ask the question 'why?' in your studies and to use the scientific method to answer it. That is, you will put forward an idea (or *hypothesis*) to try to explain what you see, and test your hypothesis with an experiment.
3	**To promote an appreciation of the importance of experimental and investigatory work in the study of biology.**	To learn to design and carry out experiments to test hypotheses; to look critically at other experiments and consider what valid conclusions can be drawn from them.
4	**To promote respect for all forms of life.**	To avoid any unnecessary damage to living things and their habitats in the study of them.

5	**To develop knowledge and understanding of fundamental biological concepts and principles.**	To study plants and animals in order to understand the general principles of biology.
6	**To develop an awareness of:** *(a)* **relationships between living organisms;** *(b)* **relationships between organisms and their environment;** *(c)* **the effect of human activities on these relationships.**	To realise that living things are dependent on each other. To discover how organisms are adapted to their environment. To study the effect of, for example, agriculture or pollution on the natural world.
7	**To develop a range of manipulative and communicative skills appropriate to the subject.**	To assemble and use apparatus and to handle biological material. To learn certain technical terms in order to understand the literature and write about what you have seen.
8	**To develop an ability to use these skills to identify and solve problems.**	To observe and describe living things; try to find out how they work and devise experiments to test your observations.
9	**To promote an awareness and appreciation of the development and significance of biology in personal, social, economic and technological contexts.**	To become aware of the impact of biology on all aspects of our lives. For example, to consider the advances of medicine in curing and preventing disease and controlling population; the use of fertilisers and pesticides in agriculture and the role of biotechnology in food and drug production.
10	**To provide:** *(a)* **a worthwhile educational experience for all, whether or not they are intending to**	To come to understand living things and the scientific method by which they are investigated, which should be useful to everybody. (The scientific method

study biology beyond GCSE level;	can be used to solve non-biological problems in everyday life.)
(b) **a suitable preparation for careers which require a knowledge of biology;**	To prepare for possible careers, such as medicine, dentistry, nursing, physiotherapy, laboratory work, agriculture, forestry, the food industry, brewing and environmental health.
(c) **a suitable foundation for further studies in biology and related disciplines.**	To build a basis for further studies in biology and related subjects, for example, the Certificate of Pre-Vocational Education, the Business and Technical Education Council examinations, or 'A' level Biology. This could lead to a diploma or degree course in Biology, Zoology, Physiology, Genetics or other related subjects.

The extent to which candidates have achieved some of the aims of the syllabus can be measured. These *objectives* fall into two groups in the National Criteria.

A KNOWLEDGE AND UNDERSTANDING

Objectives	***Practical results***
Candidates should be able to:	
1 **demonstrate knowledge and understanding of biological facts and principles, practical techniques and safety precautions;**	To recall facts and explain principles; to carry out, or describe how you would carry out, investigations safely.
2 **demonstrate knowledge and understanding of the personal, social, economic and technological applications of biology in modern society.**	To use biological facts and principles to understand the problems of society.

B SKILLS AND PROCESSES

Objectives Candidates should be able to:	*Practical results*
1 make and record accurate observations;	To describe and measure accurately and display your observations clearly.
2 plan and conduct simple experiments to test given hypotheses; *3* formulate hypotheses and design and conduct simple experiments to test them;	To devise and carry out experiments to test your own or given hypotheses.
4 make constructive criticisms of the design of experiments;	To comment on flaws in the design of experiments and suggest improvements.
5 analyse, interpret and draw inferences from a variety of forms of information, including the results of experiments;	To draw valid conclusions from data.
6 apply biological knowledge and understanding to the solution of problems including those of a personal, social, economic, and technological nature;	To use a knowledge of living things to solve the problems of society.
7 select and organise information relevant to particular ideas and communicate this information cogently in a variety of ways; *8* present biological information coherently.	To extract relevant information and present it clearly.

These overall aims and objectives are worth thinking about before you get down to the details of the subject; they suggest why you are spending time in studying this particular subject and what you can expect to gain from it.

SECTION 2

Syllabus content

Each of the five exam groups divides the content of the biology syllabus under different headings. These are different ways of arranging the content without necessarily altering the subject-matter of biology.

The syllabuses listed here are draft or approved syllabuses for the 1988 GCSE exam. Each syllabus includes the minimum core of content set out in the National Criteria. The syllabuses differ in the depth and breadth of the topics within the themes.

LEAG DRAFT SYLLABUS

Two alternative syllabuses are offered in section A, depending on the environment studied. These appear in two columns. The environment for syllabus A is a group of deciduous trees, and for syllabus B it is a fresh-water pond.

Section A: An environment and the diversity of organisms within it

Syllabus A: *A group of deciduous trees*	*OR*	*Syllabus B:* *A fresh-water pond*
1 The ecosystem		
(a) Definitions		**(a)** Definitions
(b) Component layers		**(b)** Component layers
(c) Feeding relationships: definitions; roles; food chains and webs; pyramid of biomass; population size		**(c)** Feeding relationships: definitions; roles; food chains and webs; pyramid of biomass; population size
(d) Field techniques: collecting; identifying; sampling; recording		**(d)** Field techniques: identifying; sampling; recording
2 Diversity and adaptation		
(a) Tree		**(a)** Herb

(b) Herb
(c) Alga
(d) Earthworm
(e) Snail
(f) Butterfly or moth
(g) Bird
(h) Mammal
(i) Fungi
(j) Bacteria

(b) Alga
(c) Water louse
(d) Pond snail
(e) Insect
(f) Bony fish
(g) Frog or toad
(h) Fungi
(i) Bacteria

3 Circulation of nutrients

(a) Carbon
(b) Water
(c) Nitrogen

(a) Carbon
(b) Water
(c) Nitrogen

4 Abiotic factors

(a) Light and temperature
(b) Soil

(a) Temperature
(b) Light
(c) Properties of water

5 Human interaction with the environment

(a) Pollution
(b) Conservation

(a) Pollution
(b) Conservation

Section B: The functioning of living organisms

Common to both syllabuses.

1 Cells

(a) Cell structure: plant; animal; variety of cells
(b) Movement of substances in and out of cells

2 Plant nutrition and transport

(a) Photosynthesis: significance; location; mechanics
(b) Water uptake
(c) Transpiration
(d) Transport
(e) Ion requirements
(f) Synthesis

3 Animal nutrition and transport

(a) Food classes
(b) Comparison with plants
(c) Human diet and food additives
(d) World food supplies

(e) The alimentary canal: definitions; structure; mechanical and chemical digestion; absorption; assimilation
(f) Transport: mammalian circulatory system; blood

4 Respiration

(a) Definition
(b) Aerobic
(c) Gaseous exchange: plants; animals
(d) Social and applied aspects

5 Excretion

(a) Definition
(b) Examples and origins
(c) Removal of wastes
(d) Regulation of body fluids

6 Reproduction

(a) Asexual
(b) Sexual
(c) Flowering plant
(d) Man: male and female reproductive systems; fertilisation; development; social aspects

7 Genetics

(a) Definitions
(b) Mechanism of inheritance
(c) Variation and selection

8 Response

(a) Definitions
(b) Plants
(c) Mammals: nervous system; eye; skin

MEG APPROVED SYLLABUS

The approximate teaching time which may be devoted to each theme is shown as a percentage.

Theme I: Diversity of organisms (5%)

1 Characteristics of living organisms
2 Classification of living organisms

(a) Classification system and keys
(b) Variety of organisms: viruses; bacteria; fungi; flowering plants; arthropods; vertebrates

Theme II: Relationships between organisms and with the environment (25%)

1 Fertile and infertile soils
2 Food chains and food webs

(a) Source and flow of energy within the ecosystem
(b) Food chains
(c) Food webs
(d) Biotechnology

3 Nutrient cycles: carbon; water; nitrogen
4 Population size and control
5 Parasitism
6 Effects of man on the ecosystem

(a) Agriculture
(b) Industrial pollution

7 Conservation

Theme III: Organisation and maintenance of the individual (45%)

1 The cell
2 Mitosis
3 Tissues, organs and organ systems
4 Nutrition in plants

(a) Photosynthesis
(b) Basic nutritional requirements for healthy growth in plants

5 Nutrition in mammals

(a) Classification of nutrients
(b) Diet
(c) Human alimentary canal

(i) Structure and function
(ii) Breakdown of food: physical and chemical; absorption and assimilation

6 Transport in flowering plants

(a) Water and ion uptake
(b) Transpiration
(c) Translocation

7 Transport in mammals

(a) Circulatory system
(b) Blood

8 Respiration

(a) Aerobic respiration
(b) Gaseous exchange in mammals and plants
(c) Anaerobic respiration

9 Excretion in mammals; nitrogenous excretion and other substances

10 Support, movement and locomotion

(a) Support in flowering plants
(b) Support in mammals

11 Coordination and response

(a) Coordination in plants: geotropism; phototropism
(b) Coordination in mammals

(i) Receptors: sensitivity to touch, temperature, pain; the eye
(ii) Nervous system: central nervous system; reflex arc
(iii) Hormones: e.g. insulin; adrenaline, thyroxine; sex hormones
(iv) Homeostasis

Theme IV: Development of organisms and continuity of life (25%)

1 Reproduction

(a) Asexual reproduction
(b) Sexual reproduction

(i) Sexual reproduction in flowering plants; flower structure; pollination; fertilisation
(ii) Sexual reproduction in humans: male and female reproductive systems; fertilisation, and foetal development; gestation and birth; sex hormones; biological control of population size

2 Growth and development

(a) Growth in plants
(b) Growth in humans
(c) Population size

3 Inheritance

(a) Variation
(b) Chromosomes as the site of genetic information
(c) Meiosis
(d) Monohybrid inheritance
(e) Selection

(i) Artificial selection
(ii) Natural selection

NEA APPROVED SYLLABUS

The relationship between assessment objectives and subject content is shown in brackets after each theme.

1 Relationship between organisms and with the environment (35%)

(a) Energy flow within ecosystems

  (i) Food chains and food webs
  (ii) Pyramids of numbers and biomass

(b) Natural cycles in the ecosystem
(c) Saprophytes as agents of decay and recycling
(d) Human influences in the ecosystem, including conservation, recycling and pollution
(e) Natural limiting factors to population size
(f) The world human population, its growth and control
(g) The diversity of organisms, to include as wide a variety as possible, with emphasis on those occurring locally
(h) The use of simple keys based on observable features

2 The organisation and maintenance of the individual (40%)

(a) Basic cell structure and organisation

  (i) The observable parts of cells, using a light microscope
  (ii) The specialisation of cells in relation to function
  (iii) Levels of organisation

(b) The chemistry of life

  (i) Carbohydrates (glucose, starch, glycogen and cellulose); fats; proteins and water
  (ii) Enzymes and the principle of enzyme action to show the protein nature of enzymes; enzymes as biological catalysts; enzyme specificity; factors affecting enzyme action and the technological use of enzymes

(c) Nutrition in plants

  (i) Leaf structure in relation to photosynthesis
  (ii) Experiments to investigate the requirements for photosynthesis
  (iii) The basic chemistry of photosynthesis
  (iv) Mineral requirements
  (v) Glucose as a product of photosynthesis and its uses
  (vi) Principles of techniques used to improve photosynthetic yield in greenhouse systems

(d) Nutrition in animals

(i) The significance of mechanical action in reducing the physical size of food particles
(ii) The significance of chemical digestion in the production of small soluble molecules from large food molecules
(iii) The absorption of food and water
(iv) Human dietary requirements, including carbohydrates, fats, proteins, vitamins, minerals, roughage and water; the energy content of foods
(v) The effects of food on health, to include an understanding of the concept of a balanced diet and some of the effects of malnutrition

(e) Movement of substances within organisms

(i) The principles of diffusion, osmosis, active uptake and mass flow and their significance in plants and animals
(ii) Gas exchange and ventilation
(iii) Structure and function of the mammalian heart
(iv) Structure of arteries, veins and capillaries in relation to their function
(v) The structure and functions of blood
(vi) Water movement in plants

(f) Energy release

(i) Aerobic respiration as the release of energy via the oxidation of glucose
(ii) Anaerobic respiration compared with aerobic respiration
(iii) Social and technological significance of respiration

(g) Homeostasis

(i) Temperature regulation
(ii) The liver as a homeostatic organ
(iii) The kidney as a homeostatic organ

(h) Sensitivity, coordination and response

(i) Phototropism as a plant growth response
(ii) Nervous control in animals
(iii) Hormonal control in animals
(iv) Effector systems

3 Development of organisms and the continuity of life (25%)

(a) Reproduction

(i) Asexual

(ii) Sexual in plants and mammals: structure; fertilisation; birth; contraception in mammals
(iii) Artificial propagation

(b) Growth

(i) Methods of measuring growth
(ii) Factors affecting the growth of animals and germination of flowering plants

(c) Variation and inheritance

(i) Continuous and discontinuous variation
(ii) Genes and chromosomes
(iii) Genes and inheritance
(iv) Causes of variation

(d) Selection

(i) Principles
(ii) Natural selection
(iii) Artificial selection

SEG APPROVED SYLLABUS

The relationship between assessment objectives and subject content is shown in brackets after each theme.

Theme I: Diversity of organisms (5%)

1 Increase in complexity
2 Diversity of organisms, local organisms may be used
3 Keys
4 Binomial nomenclature

Theme II: Relationship between organisms and between organisms and the environment (35%)

1 Ecology
2 Food chains and food webs; concepts of energy flow; pyramid of biomass
3 Population size and its control
4 Ecological interdependence of organisms
5 Cycles of matter: carbon; nitrogen; water
6 Human influences: land use; use of coal and oil; pollution; pest control; recycling

Theme III: Organisation and maintenance of the individual (40%)

1 The cell

2 Principles of enzyme action and diffusion; control of internal concentration by cell wall; contractile vacuole; kidneys; food preservation

3 Life processes

(a) Photosynthesis: the process; products; the leaf as the site of photosynthesis

(b) Nutrition: basic requirements

(i) Plant nutrition: mineral nutrition; use of fertilisers

(ii) Animal nutrition, e.g. humans; diet and health

(c) Digestion: in plants; in animals, e.g. humans; anatomy of digestive system; enzymes and digestion; oral hygiene

(d) Transport of soluble substances within organisms: in plants; by the blood in humans; control of bleeding

(e) Respiration: process; aerobic and anaerobic; principles of gaseous exchange; human breathing system; gaseous exchange in flowering plants; health aspects, e.g. smoking, air pollutants; mouth-to-mouth ventilation

(f) Excretion: removal of waste products of metabolism; origin of excretory products; urinary excretion in mammals; excretion in flowering plants; toxic materials from the environment; health aspects, e.g. kidney failure and pollen allergies

(g) Sensitivity and response: phototropic responses in plants; sensitivity of the skin; outline of sense organs; principles of stimulus and response; reflex actions; the brain; drugs

(h) Hormonal control: sex hormones; plant hormones; artificial use of hormones

(i) Homeostasis: liver; skin; kidneys

(j) Support and movement: skeletons; movement and principles of leverage

Theme IV: Development of organisms and the continuity of life (20%)

1 Life as a cyclic process

2 Reproduction

(a) Asexual and sexual

(b) Human reproduction: male and female reproductive systems; fertilisation; birth; sexually transmitted diseases; biological principles of contraception

(c) Reproduction in flowering plants: structure; pollination; fertilisation; germination

3 Growth

4 Genetics

(a) Chromosomes; cell division; reduction division; significance of chromosome pairs; sex chromosomes
(b) Genes
(c) Genetic inheritance; monohybrid cross
(d) Genetic engineering
(e) Variation and its significance; genetic variation; environmental variation

WJEC APPROVED SYLLABUS

Theme 1: Diversity of organisms

1 Living organisms
2 Plant and animal cells

(a) Plant
(b) Animal

3 The variety of life

(a) Algae
(b) Mosses
(c) Ferns
(d) Flowering plants

(i) Herbaceous
(ii) Woody

(e) Fungi
(f) Protozoa
(g) Annelids
(h) Arthropods
(i) Molluscs
(j) Fish
(k) Birds
(l) Mammals

Theme 2: Organisation and maintenance of the individual

1 Plant nutrition and transport

(a) Photosynthesis
(b) Water uptake
(c) Transport and transpiration

2 Animal nutrition and transport

(a) Food classes
(b) A brief introduction to human diet
(c) The alimentary canal

(i) Basic functions
(ii) Structure
(iii) Digestion
(iv) Absorption and assimilation

(d) Transport

(i) Blood
(ii) Circulation of the blood

3 Respiration

(a) Gaseous exchange

(i) Plants
(ii) Mammals

(b) Release of energy

(i) Aerobic
(ii) Anaerobic

4 Excretion

(a) Examples and origins
(b) Regulation of body fluids and removal of wastes

5 Sensitivity in plants and animals

(a) Plants
(b) Mammals

(i) Eye
(ii) Nervous system
(iii) Movement
(iv) Endocrine system

Theme 3: Development of organisms and continuity of life

1 Reproduction

(a) Flowering plant
(b) Man

(i) Male reproductive system
(ii) Female reproductive system
(iii) Fertilisation and development
(iv) Social aspects

2 Cell division and genetics

(a) Cell cycle and growth
(b) Inheritance, variation and selection

Theme 4: Relationships between organisms and with the environment

1 Energy flow within ecosystems
2 Population size and control
3 Cycles of matter
4 Human interactions with the environment

(a) Pollution
(b) Conservation

5 Parasites as pathogens

SECTION 3

Preparing for the exam

Success in an exam depends on the organisation of a study schedule and revision programme. In your study schedule, your aim should be to understand each topic in the syllabus thoroughly. Let's see how this can best be done.

YOUR STUDY SCHEDULE

Practical work

One of the chief differences between the old 'O' level/CSE exams and the GCSE is that there is now more emphasis on practical work. This is welcome, as our knowledge of living things derives from observations and experiments. You will spend more time making your own observations and carrying out your own experiments than did the 'O' level/ CSE students.

Finding out for yourself is probably the most effective way of learning, as personal involvement will help you to understand and to remember. How do you set about it?

Your teacher will guide you in using scientific methods. Suppose you are studying how plants feed; you will first be encouraged to observe. You will not just use your eyes, but listen, touch and smell. The more senses you use, the easier you will find it to remember. You will notice that pondweed bubbles in the sun, but that the bubbles have no smell.

You will then be encouraged to suggest ideas, or hypotheses, to explain what you have seen. You will learn to state a hypothesis as clearly and concisely as you can. For example, you may suggest that pondweed gives off oxygen in the light. In which case you must devise an experiment to test your hypothesis. Join in a class discussion on how to do this. Some of your ideas may be discarded as unworkable, but don't be put off. You will learn by making mistakes.

When you have decided on a suitable design, you will assemble the apparatus and carry out the experiment. Think about the reasons for using any particular chemicals or procedure; about the precautions you must take; about any safety measures that are necessary and about the readings you must take. Record all these points when you write notes on your method, and draw a labelled diagram of your apparatus to help you remember what you did.

Record the results of the experiment clearly. This should be done in a table if you have taken several readings. The relationship between two sets of measurements may be shown by plotting a graph from your results.

The final step is to draw a valid conclusion from the results. Consider the aim of the experiment and state whether or not the results support the hypothesis.

As well as carrying out experiments, you will be trained to observe and draw specimens. Use a hand lens to see as much detail as you can in larger organisms. Cells or small organisms can only be seen under a microscope. You will prepare slides and look at them under the microscope. You will need to practise finding specimens and focusing on them until you are confident in using this instrument.

As practical work takes up so much time on the course and is considered so important in the GCSE, it must be assessed.

At least 20% of your final exam mark will be allocated to practical skills. During the course, your teacher will note which skills you have acquired. This will be reported to the examining group at the end of the course. The group will decide on the actual percentage of marks to be awarded to you for your practical work.

As there is a continual assessment of practical work, it is important to develop a sound technique from the start. Let's see which practical skills may be assessed and how to perfect them.

Practical skills assessed

Observation

In observation, the more details you notice the better. The average student will note the number of petals on a flower, but the good student will note the exact shape of a petal and any markings on it.

The structures of organisms and apparatus are best recorded by diagrams, but don't worry if you are not artistic. The best biological drawings are simple but accurate line diagrams, labelled to show as much detail as possible.

Handling material and apparatus

Before you begin to assemble apparatus, think carefully about the aim of the experiment. Put all the apparatus you will need on your bench and assemble it methodically following the instructions or diagrams. Check that corks and seals are airtight and that the apparatus is clamped in the correct position. Take safety precautions in all your experiments.

Practise using the microscope and preparing microscope slides. Make sure no material is protruding from the cover clip. There should be no air bubbles, but not too much mounting fluid.

Measuring

Take readings and measurements with care for each experiment. How accurately can you measure the volume of water in the measuring cylinder in the diagram? Now estimate the diameter of the cell. (The actual measurements are given at the bottom of the next page.)

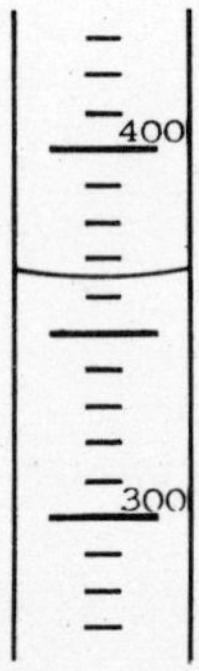

Part of a measuring cylinder in cm^3

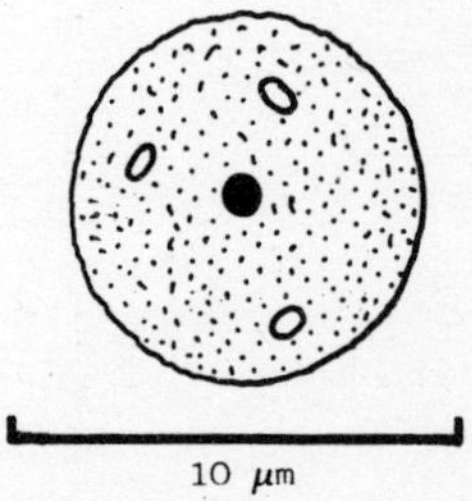

A cell

Recording

Record your readings in a table. This may already be prepared, or you may be asked to construct your own. Each column should be headed clearly and the units stated, as in the example which follows. Record every reading you take and enter it directly in the table, not on a scrap of paper or the back of your hand!

Table of results of experiment to investigate the effect of temperature on the bubbling rate of yeast

Temperature in °C	**10°**	**20°**	**30°**	**40°**
Number of bubbles in 1st minute	0	2	4	9
Number of bubbles in 2nd minute	1	3	6	12
Number of bubbles in 3rd minute	2	2	5	11
Average number of bubbles/min.	1	2	5	11

Plot a graph to show the relationship between two measurements. The graph should have a heading; the measurements that you, as the experimenter, have selected should be on the horizontal axis; the axes should be labelled with the units; the largest possible scale used; the points accurately plotted and the graph drawn, as in the example here.

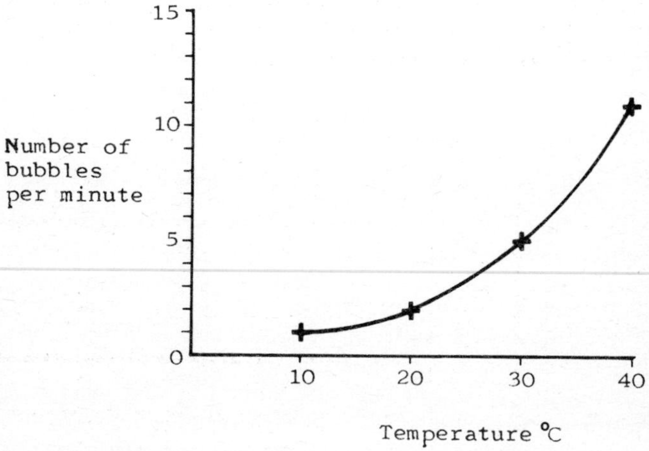

Graph of bubbling rate of yeast against temperature

Data and its interpretation

In interpreting your results, make sure that your conclusions are based only on the results of the experiment, not on what you think should have happened or what you have read about. For example, from the results in the table, you would conclude that the bubbling rate of yeast increases with temperature up to 40°C.

Arithmetical calculations, such as addition, subtraction, averages and percentages, will have to be done in some of your assessments.

Design

You may be asked to criticise the design of an experiment. This means you comment on basic flaws, such as the lack of a control, not on a fault in the apparatus, such as an ill-fitting cork.

Answers (Measuring): volume of water = 365 cm^3; diameter of cell = 7.5 μm

You could improve the design of the bubbling yeast experiment by repeating it with dead yeast as a control, or collecting the gas and measuring its volume.

Remember these points if you are asked to design your own experiment. Allow for the variation in living things by repeating the experiment several times and taking an average reading.

Formulate a hypothesis to explain your observations. You will have plenty of experience of this during the course.

So, as you see, carrying out experiments requires a great deal of thought as well as manual dexterity. The experience will add to your knowledge more effectively than any other method, but unfortunately there is not enough time to learn everything in this way.

Your teacher will demonstrate other experiments to you and explain certain topics, which you will make notes on.

Note taking is an art worth mastering. So let's consider how to set about it.

Note taking

'Chalk and talk' may still be the best way of explaining, but your school will probably have a battery of visual aids to help you to grasp an idea. Slides, film loops, videos, models and charts are all useful, since they will help you to build up patterns in your mind, which is what learning is all about.

You will be encouraged to discuss and question. The more you get involved in discussion, the clearer the ideas will become to you.

Thorough notes are essential to help you remember and revise what you have learnt. Underline the headings; main headings with a double line and side headings with a single line. Key words should also be underlined.

Do not write down any technical term for the first time without explaining its meaning. Perhaps you could keep a glossary of these terms in the back of your book. It may seem boring to have to learn all these words, but you will need them in order to understand the text-books and to describe your own experiments.

Include flow charts in your notes where possible. A flow chart is a useful way to summarise a process by connecting words by arrows, as in the example opposite. You will find that these charts are much easier to remember than lists of words.

Illustrate any structure with a labelled diagram. Always write a heading for the drawing, explaining what it is and which view you are looking at, e.g. transverse section of a holly leaf. Again these notes will help to form patterns in your mind which will stick.

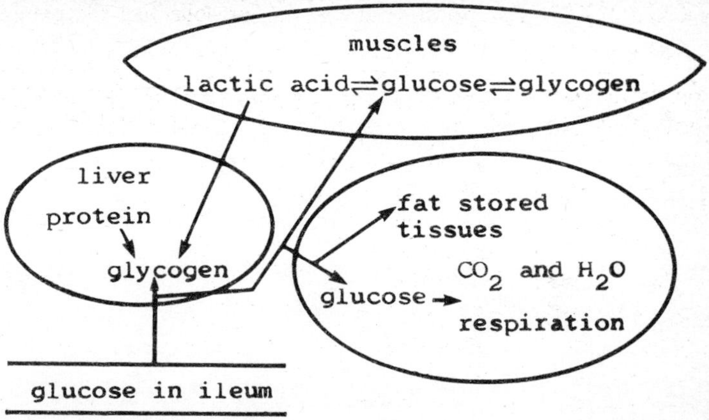

Flow chart to show carbohydrate metabolism

Leave a margin so that you can add to your notes from time to time, and above all write legibly. A hurried scrawl may make sense today, but will you decipher it several months later?

It's a good idea to conclude each topic with a summary. This should be a paragraph with the key words underlined.

Reviewing

After you have finished a topic, you may find that, two days later, you can remember very few of the facts. This can be put right by 'reviewing': a quick re-read of your notes and summaries. If they are well organised, you will find that far more of the facts stick this time. Repeat the method a few days later, and the results will be even better.

Reading

Your experimental work and class notes must be consolidated by reading. The teacher will give you references for each topic, or you can find them yourself in the index of your text-book.

Try to read a topic before you discuss it in class, so that you can take more part in class discussion. Read the topic again after you have taken your class notes, and add any further points in the margin. Jot down anything you don't understand and ask at the next lesson.

You may wonder where you are going to find the time for all this, but, with practice, you will be able to improve your reading speed quite dramatically. Instead of reading one word at a time, focus on the whole phrase at once. This is not the same as skimming through a page, when you miss several sentences out. Try to take in every word, not only increasing your reading speed, but also your understanding of the passage.

In time, you will find you can more than triple your reading speed.

Essay writing

Experimental work, class discussion and reading should all help you to understand your subject. Essay writing is your chance to show how well you have understood it. Essays can also help your teacher to detect any misunderstandings and sort them out for you.

Before you begin writing, make a plan of your essay. Think about the title carefully. 'Describe' means that you should give a full account, but 'discuss' requires a more analytical approach, providing arguments for and against.

Plan an introduction which will suggest the line the essay will follow. Jot down the major points you are going to make and arrange them in a logical sequence. Illustrate any point or principle with specific examples, and round off your essay with a summary or conclusion.

With your study schedule completed, you will be ready to tackle the next stage - revision.

Many students sail through their course and then disappoint themselves in the exam. Revision seems to be the problem, so let's see how to solve it.

ORGANISING YOUR REVISION

Revision timetable

Before you actually start your revision, you need to make a timetable. Begin by drawing up a checklist of topics under the main headings of the themes for your examining group. Refer to the detailed list of the subject content of your syllabus in section 2 to make sure that you don't leave any topics out. For the MEG, for instance, the themes and a first topic would be as shown opposite.

When you have drawn up your checklist, the next step is to decide the order in which to revise the topics. If you are attending classes and the class is following a detailed revision plan, then take this into consideration when drawing up your own plan. Otherwise, draw up a plan to suit your own individual needs, taking into account your strengths and weaknesses and how much time you feel you need to spend on

Themes	Topics
Diversity of organisms	Characteristics of living organisms; classification and keys; etc.
Relationships between organisms and with the environment	
Organisation and maintenance of the individual	
Development of organisms and the continuity of life	

each topic. Allow for the percentage weighting of each theme if this is shown in brackets after the headings in the syllabus section.

The best way to organise your revision is by dealing with all the topics under each of the four main themes in turn, rather than by switching from one topic to another, regardless of the themes to which the topics are related. Leave plenty of time for your revision by starting early.

Here is an example of a twelve-week revision plan drawn up by a student preparing for the MEG exam:

Week	Topics
1	*Diversity of organisms. Relationships between organisms and environment:* soils; food chains and webs;
2	biotechnology; nutrient cycles; population size and control;
3	parasitism; agriculture; pollution; conservation; ecological study.
4	*Organisation and maintenance of the individual:* cells and cell division; tissues, organs, systems;
5	nutrition in plants;
6	nutrition in mammals;
7	transport in flowering plants; transport in mammals;
8	respiration; excretion in mammals;
9	support in plants; movement and support in animals; coordination in plants; coordination in mammals.
10	*Development of organisms and continuity of life:* reproduction in plants and mammals; growth and development in plants and humans;
11	inheritance; selection.
12	Review. Try past papers.

Before you make your own plan, you must decide how many hours you can spend a week on revision. Be realistic here, taking into account the number of subjects you are studying, how much of the syllabus has still to be covered and what other commitments you have. Your revision will be much more efficient if you allow yourself a reasonable amount of time for leisure as well.

For instance, if you are taking six subjects, you may decide that each needs an average of three hours a week in a twelve-week programme. For the first four weeks this may only be two hours, as you must leave some time to finish the coursework. This could be stepped up to three hours in the second four weeks and four hours in the final four weeks, when part of the revision time might be during school hours.

It is worth checking with your teacher about when the course is likely to be completed and what the revision plans are, so that you can coordinate your own plan.

The revision timetable given here has been worked out for someone taking six subjects: English, maths, French, history, chemistry and biology. Three hours have been allowed for each subject, giving a total of eighteen hours revision time per week.

DAY	9.00-5.00	5.00-6.00		7.00-8.00	8.00-9.30
Monday	school	maths	D	English	chemistry
Tuesday	school	English	I	French	history
Wednesday	school	French	N	maths	chemistry
Thursday	school	English	N	maths	biology
Friday	school	French	E	free	free
Saturday	9.00-10.30, 10.30-12.00, 12.00-5.00 history / biology, free	free	R	free	free
Sunday	free	free		free	free

A revision timetable

You may decide you would rather work every night and have Saturday completely free. Adjust the times as you like, but make sure you get the hours in.

If you are taking eight subjects altogether, you may decide to allow only two and a half hours per subject and either work half an hour longer each evening or put in an extra two hours at the weekend.

When you have made a realistic plan, you must stick to it. This is the key to success. For instance, if you want to go out one Wednesday evening, allow for this by doing Wednesday's work in advance the week before, but never during the following week. It's not a bad idea to get ahead anyway sometimes, so that you always feel you are on top of the schedule, rather than trying to keep up with it. Psychology plays a big part in revision!

Now that you have decided when you are going to revise a particular topic and how much time is available, let's see how you can make best use of this time.

Remembering facts

First find a suitable place to revise in; preferably somewhere quiet, warm and comfortable.

Find what your concentration time is by timing yourself reading part of a text-book which you haven't read before, trying hard to take it in. Note the time at which your mind begins to wander. This is the end of your concentration time, so rest for five minutes.

Plan your revision around your concentration time, which will probably be about thirty minutes, with five-minute breaks. Decide how much you can realistically learn in that time and learn it. You will feel a sense of satisfaction then in having set out with an aim and having achieved it. Trying to revise an indefinite amount of work at one time is excessive, and may lead to your abandoning the whole programme as hopeless.

As you read through your notes, write down key words and join them up into pattern diagrams to help you remember them.

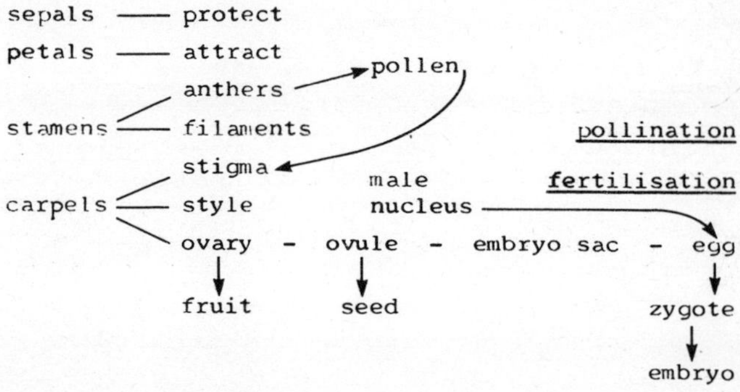

Pattern diagram of sexual reproduction in a flower

At the end of your study session, review the topic by reading through the key words. Repeat this a few days later, and you will find that you have remembered most of the facts.

Memory aids

Any device to help the memory is known as a *mnemonic*. There are several different kinds of mnemonics.

Picture association is useful in linking words. For instance, to remember that Vitamin A is found in carrots and prevents night blindness, picture an eye seeing a single carrot in the dark.

Initial letters, or some other part of a technical word, can be taken and put into everyday words to remember a list or sequence. For example, in the nitrogen cycle, 'urea – ammonia – nitrites – nitrates – amino acids – proteins' can be remembered by: 'u at night ate 'am and peas'. The more ridiculous mnemonics are, the easier it is to remember them.

Rhymes are useful for remembering facts and figures, for example, the spelling rule: '*I* before *E* except after *C*.' You can probably make some up for yourself.

Although these aids help your memory, they do not help your understanding. So, when you have thoroughly revised a topic, test your understanding by tackling specimen questions. You will find some in the next section. Write down your answers and check them against the types of answers given in section 5.

The last week of your plan should be given to reviewing the whole syllabus and testing yourself with these papers.

SECTION 4

Sample questions

The following questions are a selection from the latest draft and approved versions that were available before this book went to press.

NEA SPECIMEN PAPERS

Paper I (2 hours)
(Taken by all candidates.)

Section A

10 In the food chain:

oakleaf → aphid → ladybird

the ladybird is feeding as a:

A decomposer
B secondary consumer
C primary consumer
D secondary carnivore *(1 mark)*

14 Which one of the graphs below shows what has happened to the human population of the world over the last 500 years? *(1 mark)*

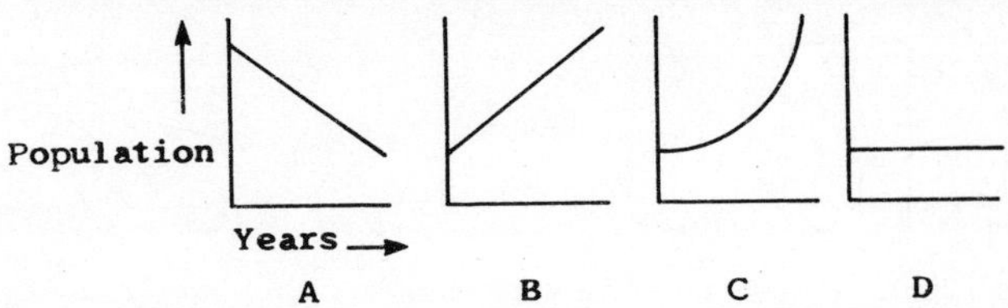

15 Which one of the following features is most likely to be found in a wind pollinated flower?

A brightly coloured petals
B production of nectar

C large, branched or feathery stigmas
D production of a scent *(1 mark)*

16 If the body cells of a fish contain 24 chromosomes, the sperm of this fish will contain:

A 24 chromosomes
B 48 chromosomes
C 12 chromosomes
D 6 chromosomes *(1 mark)*

17 If B represents the dominant eye colour brown and b represents the recessive eye colour blue, the genotypes of the brown-eyed parents who produced a blue-eyed child would be:

A BB and bb
B BB and Bb
C bb and bb
D Bb and Bb *(1 mark)*

Matching pairs questions

In questions 25–40 each group of questions has a set of responses, ***A***, ***B***, ***C*** and ***D***. In each group each letter may be used once, more than once, or not at all. For each question, select the *best* response and mark its letter on the answer sheet.

Questions 25–28

The diagram below illustrates an experiment to test a leaf disc for starch. Discs were transferred in the directions shown by the arrows.

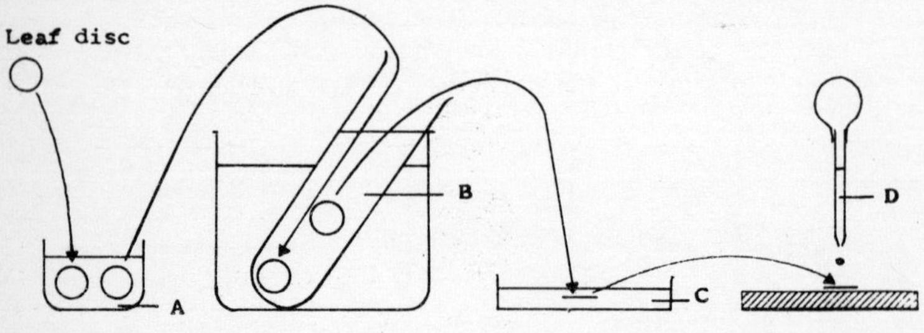

Which of the liquids ***A*** to ***D*** is . . .

25 iodine in potassium iodide solution? *(1 mark)*
26 ethanol (alcohol)? *(1 mark)*
27 used to soften the discs? *(1 mark)*
28 used to decolourise the discs? *(1 mark)*

Section B

6 The diagrams in the table below show four different kinds of woodlouse.

Examine the diagrams carefully, then describe *one* way in which each kind of woodlouse differs from the other three. *Ignore differences in size.* *(4 marks)*

Woodlouse			
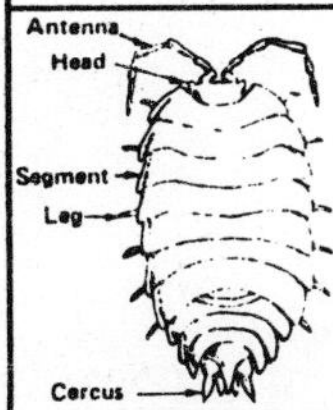	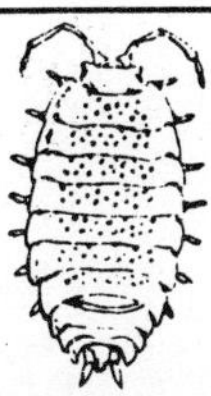	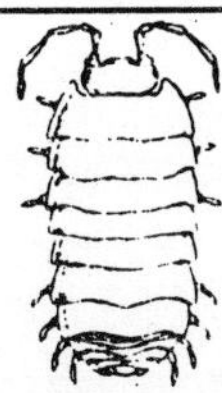	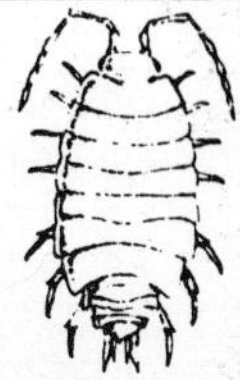
Oniscus asellus	*Porcellio scaber*	*Armadillidium vulgare*	*Philoscia muscorum*

7 There are two types of clover plant. One type makes cyanide in its leaves, the other type cannot make cyanide. Cyanide is poisonous to most organisms, but not to the clover plants which make it. The clover plants growing on a mountain were studied. The chart shows the proportions of the two types of clover plants at different heights.

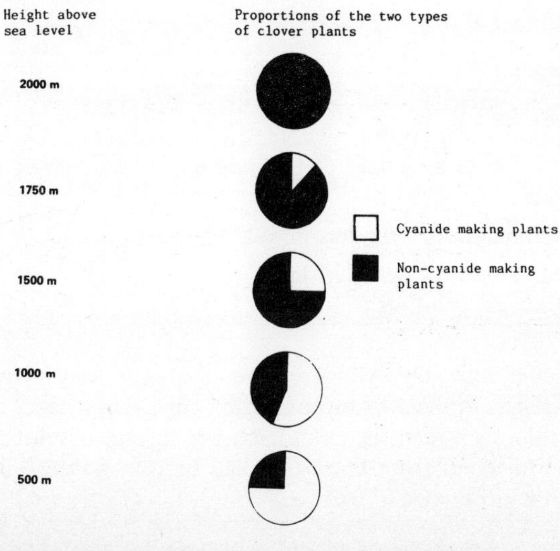

(*a*) Suggest *one* reason why cyanide in the leaves might help clover plants to survive. *(1 mark)*

(*b*) Suggest why the proportions of the two types of plant change as we go up the mountain. *(2 marks)*

9 In an osmosis investigation using potato, six cylinders of potato were cut to a length of *exactly* 50 mm.

Three of the cylinders were put in water.

Three of the cylinders were put in 30% sugar solution.

After two hours the cylinders were taken out and dried. They were put on 2 mm graph paper to measure their lengths. The outline of each cylinder is drawn on the graph paper below.

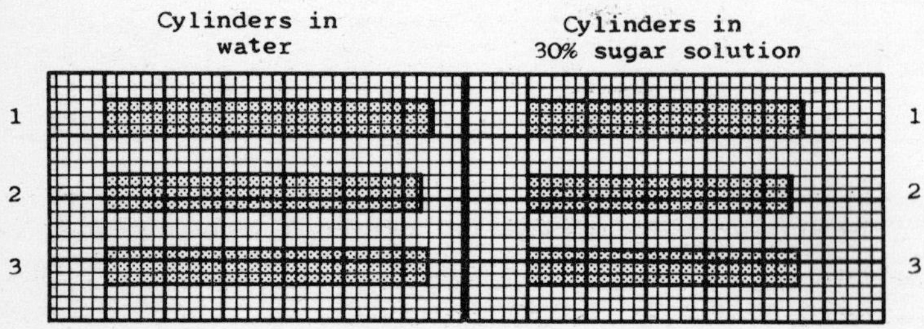

(*a*) Find the lengths of the three cylinders from each of the liquids and record your results in the table below. *(2 marks)*

	Water	*30% sugar solution*
Cylinder 1		
Cylinder 2		
Cylinder 3		

(*b*) Find the mean length of the cylinders in each liquid after two hours.

Mean length of cylinders in water = ...

Mean length of cylinders in 30% sugar solution = ...

(2 marks)

(*c*) At the start, the cylinders were all 50 mm long. Why did the cylinders in the 30% sugar solution change in length? *(1 mark)*

(*d*) After two hours, the cylinders from the sugar solution would feel different from those that had been in water. What would be the difference in feel? *(1 mark)*

12 A pupil was asked to compare the energy values of sugar and butter. He set up the apparatus shown in the diagram below.

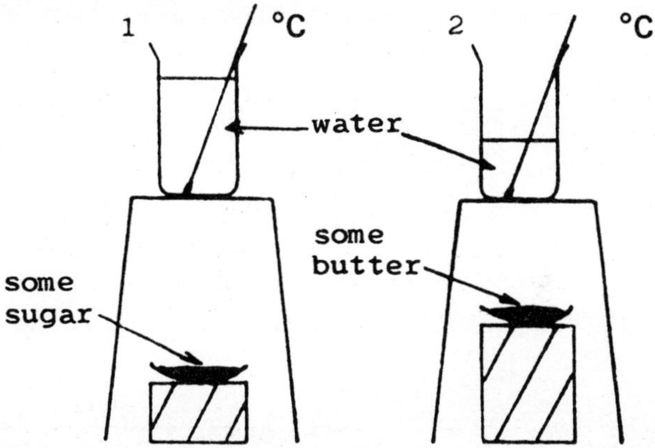

Suggest *three* improvements he could make to his apparatus so that he could compare the energy values accurately. *(3 marks)*

14 The table below shows the winning times in races held at three Olympic Games at three different cities.

Race	Tokyo – 1964 200 m above sea level	Mexico – 1968 2,240 m above sea level	Munich – 1972 52 m above sea level
100 m	10.0 s	9.9 s	10.2 s
200 m	20.3 s	19.8 s	20.0 s
400 m	45.1 s	43.8 s	44.7 s
800 m	1 min 45.1 s	1 min 44.3 s	1 min 45.9 s
1,500 m	3 min 38.1 s	3 min 34.9 s	3 min 36.3 s
5,000 m	13 min 48.8 s	14 min 28.4 s	13 min 26.4 s
10,000 m	28 min 24.4 s	29 min 27.4 s	27 min 38.4 s

(a) Using only the information in the table, what are the effects of height above sea level on performance of the athletes? *(2 marks)*

(b) State *two* pieces of evidence which support your answers to section ***(a)***. *(2 marks)*

Paper II Level P (1½ hours)

(Designed for grades G to C.)

2 *(a)* The graph below shows the amount of smoke in the air around a city.

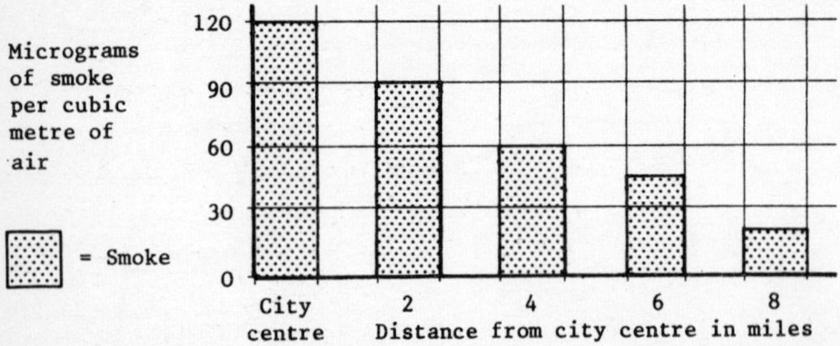

Suggest one reason why the amount of smoke in the air changes as you move away from the city centre. *(1 mark)*

(b) The graph below shows the deaths of men from bronchitis around the *same* city.

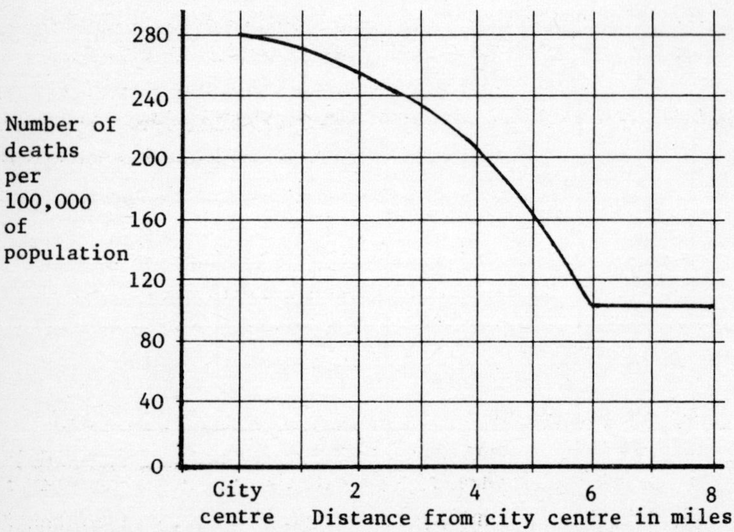

Suggest why the number of deaths from bronchitis decreases as you move away from the city centre. *(1 mark)*

(c) The city council would like to cut down the deaths from bronchitis. Suggest *two different* things which they could do. *(2 marks)*

7 The diagram below shows part of an artificial kidney machine.

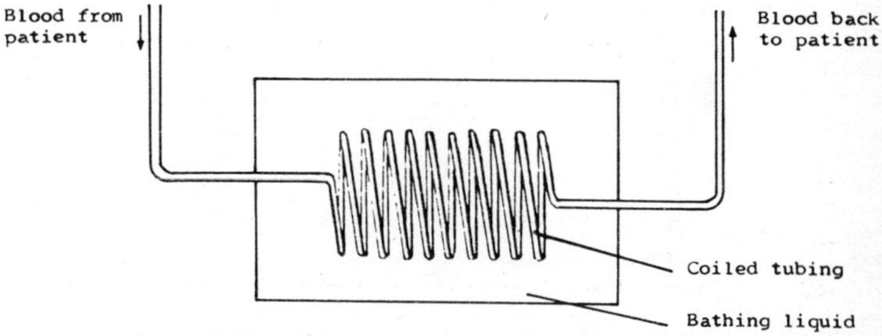

(a) *(i)* Explain why the tubing is coiled rather than straight. *(2 marks)*

(ii) Suggest a suitable material for the coiled tubing. *(1 mark)*

(iii) Explain why urea passes from the blood to the bathing liquid. *(2 marks)*

(b) State one advantage of having a kidney transplant rather than frequent treatment with an artificial kidney machine. *(2 marks)*

8 The diagram below shows a section through the elbow.

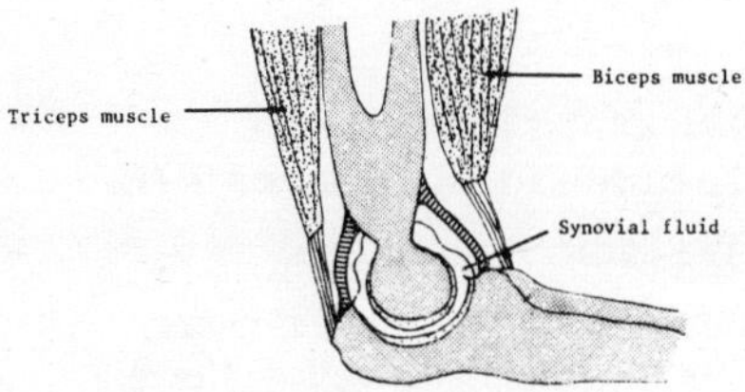

(a) *On the diagram:*

(i) label a tendon with the letter *X*; *(1 mark)*

(ii) label a ligament with the letter *Y*. *(1 mark)*

(b) Why does the joint have synovial fluid? *(1 mark)*

(c) Explain why we need a triceps muscle as well as a biceps muscle. *(3 marks)*

12 A gardener has three plants of the same kind.

Plant *A* has red flowers.
Plant *B* has red flowers.
Plant *C* has yellow flowers.

The gardener carried out the crosses shown below.

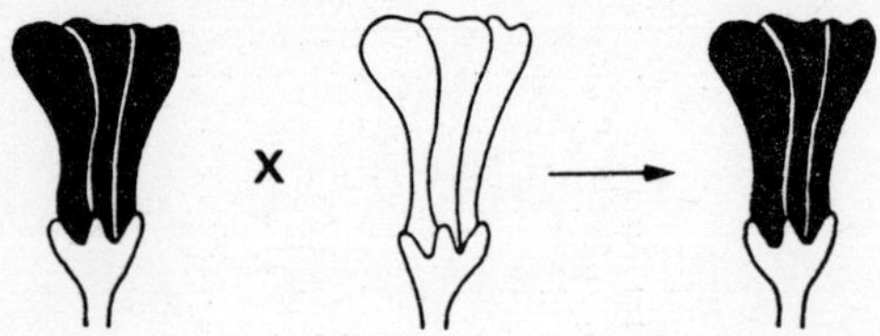

Plant A (red flowers) crossed with Plant C (yellow flowers) — All plants produced have red flowers

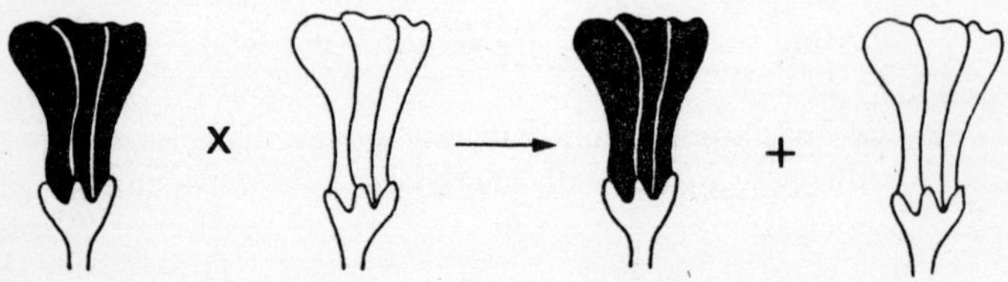

Plant B (red flowers) crossed with Plant C (yellow flowers) — Half the plants produced have red flowers and half have yellow flowers

(a) *(i)* Which flower colour is dominant? *(1 mark)*
(ii) Explain the reason for your answer to *(i)*. *(1 mark)*

(b) Explain what the gardener would do

(i) to get plants which all have red flowers; *(1 mark)*
(ii) to get plants which all have yellow flowers. *(1 mark)*

Paper II Level Q ($1\frac{1}{2}$ hours)

(Designed for grades E to A.)

2 The next graph shows the relationship between respiratory tract diseases and air pollution along a straight line from a rural area into a city centre.

(a) *(i)* Suggest one factor which contributes to the change in smoke concentration from the rural area to the city centre. *(1 mark)*

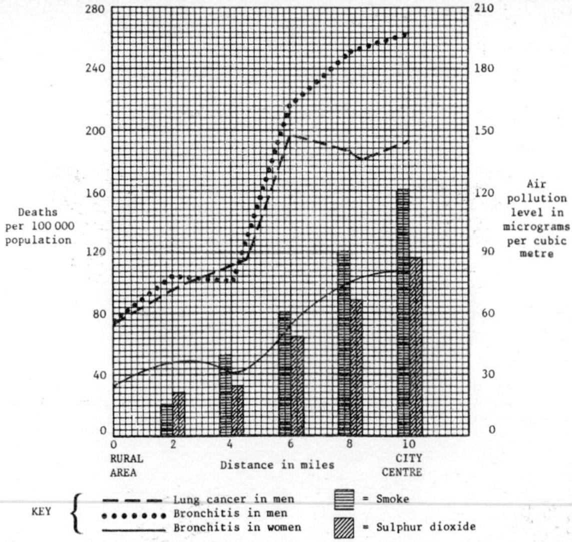

(ii) Suggest a different factor which contributes to the change in sulphur dioxide from the rural area to the city centre. *(1 mark)*

(b) Suggest a cause for the difference in the death rates from bronchitis between men and women. *(1 mark)*

(c) The graph below shows the changes in the lichen cover on walls from the rural area to the same city centre. Lichens are simple plants.

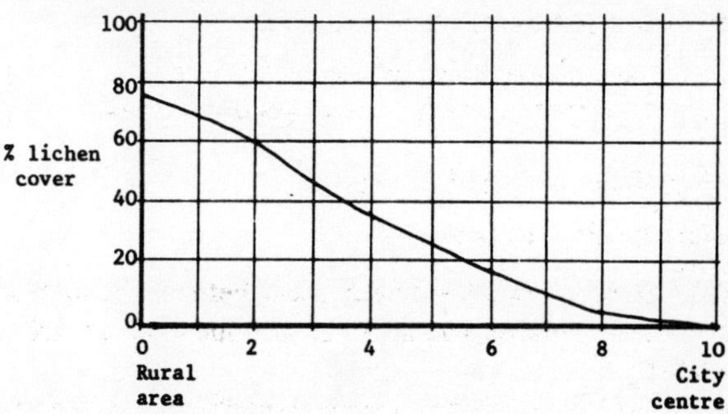

Using the information from the *two* graphs, suggest a relationship between air pollution and lichen cover.

(1 mark)

(d) What changes might enable lichens to recolonise the city centre *without* their being artificially introduced? *(2 marks)*

3 Write an account of man's effects on *either* an aquatic *or* a land habitat which you have studied. *(8 marks)*

6 The diagram below shows part of a vertical section through part of a green leaf.

(a) *(i)* Give the main advantage to the plant of having large air spaces as shown in the diagram. *(1 mark)*

(ii) Suggest one benefit to the plant of having the palisade mesophyll elongated as shown in the diagram. *(1 mark)*

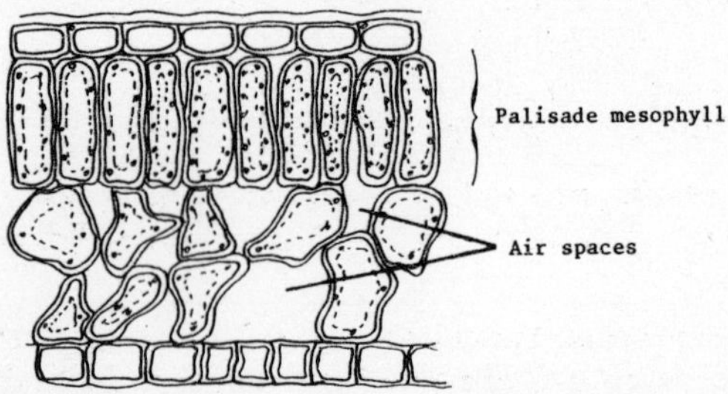

(b) Increasing both the carbon dioxide content of air and the temperature will give an increased yield of crops grown in a greenhouse.

(i) Explain how a grower could increase the carbon dioxide content of the air with the same equipment used for heating the greenhouse. *(2 marks)*

(ii) How would a grower decide on how much heat to provide in a greenhouse? *(2 marks)*

10 The diagram below shows a section through the human eye.

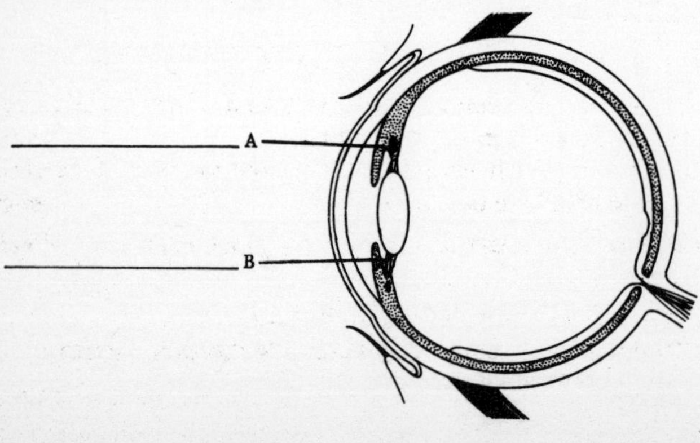

(*a*) *On the diagram*, label the structures marked *A* and *B*.

(2 marks)

(*b*) The diagrams below show a mammalian eye as seen from the front.

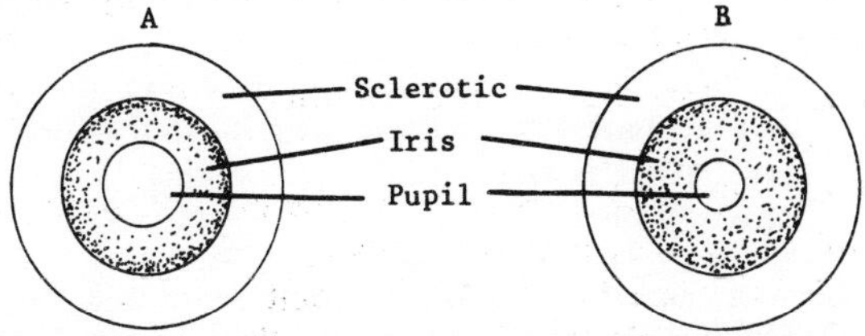

(*i*) What environmental change would cause the alteration from *A* to *B*? *(1 mark)*

(*ii*) Explain fully how the nervous system and muscles bring about this alteration. *(5 marks)*

12 (*a*) Describe the events which occur after the release of sperms into the vagina of a human female until implantation of the embryo. *(4 marks)*

(*b*) The chart below shows how the thickness of the uterus lining and the levels of two hormones *A* and *B*, made in the ovaries, vary during the menstrual cycle. The first month is a normal menstrual cycle but fertilisation occurs during the second month.

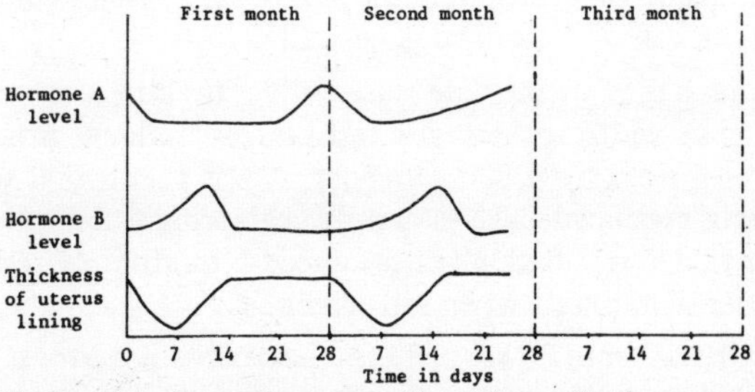

(*i*) *On the chart above*, mark, with an *X*, the most likely time of ovulation in the first month. *(1 mark)*

(*ii*) During which period in the second month is fertilisation likely to have occurred? *(1 mark)*

(*c*) *Complete the chart above* to show what happens, following fertilisation, to:

(*i*) the uterus lining in the third month;

(*ii*) the levels of hormones *A* and *B* in the third month.

(3 marks)

(d) One type of contraceptive pill contains a mixture of hormones *A* and *B*.

(i) Explain briefly how this pill works as a contraceptive. *(1 mark)*

(ii) If she is using the contraceptive pill, it is usual for a woman to take a hormone pill each day for 21 days and then to take a pill without any hormones for the next 7 days. What is the advantage of taking the hormones for 21 days only? *(1 mark)*

13 When the mechanism of inheritance of flower colour in garden peas was investigated, red-flowered plants were crossed with white-flowered plants. The first generation of plants all had red flowers. However, when these red-flowered plants were allowed to self-fertilise, about 25% of the offspring had white flowers, the remainder having red flowers.

In a similar investigation with snapdragon plants, when red-flowered plants were crossed with white-flowered plants, the resulting first generation all had pink flowers. When these pink-flowered plants were self-fertilised, 25% of the offspring had white flowers, 25% had red flowers and 50% had pink flowers.

(a) Suggest why the results obtained with the garden pea are different from those obtained with the snapdragon plants. *(1 mark)*

(b) By means of a diagram, show how the results for the snapdragon can be explained genetically. *(4 marks)*

(c) Some barley plants are susceptible to attack by mildew (a fungus) whilst others are resistant to mildew attack. In an investigation by a plant breeder, it was found that susceptible plants produced only susceptible offspring when self-fertilised, but that a resistant plant produced a mixture of resistant and susceptible plants when self-fertilised.

(i) How would the plant breeder obtain a stock of barley plants which were all resistant to mildew? *(2 marks)*

(ii) Assuming that resistance to mildew is controlled by a single gene, what must be the genotype of the resistant stock? *(1 mark)*

15 The graphs which follow show the patterns of survival in the brown trout and man.

(a) Give two reasons why the pattern of survival for man is different from that for the brown trout. *(2 marks)* *(b)* Explain how the pattern in the graph for the brown trout would support Darwin's theory of Natural Selection. *(3 marks)*

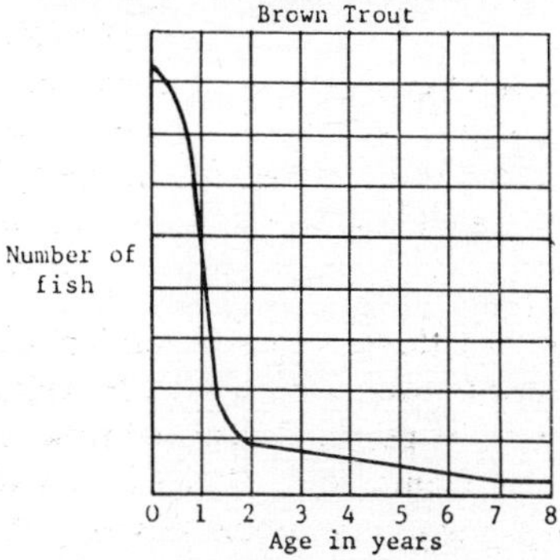

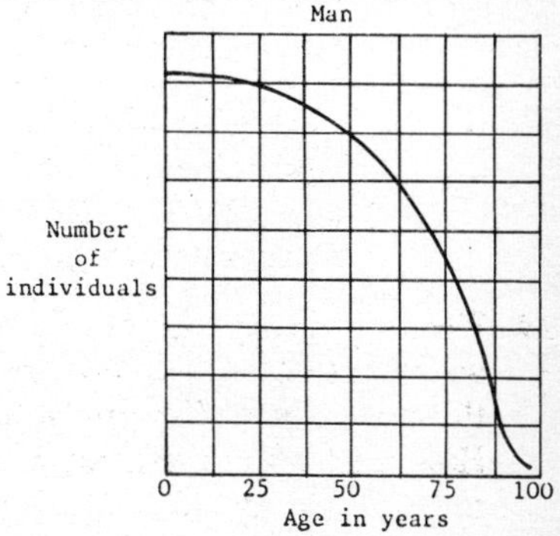

SEG SPECIMEN PAPERS

Paper 1 (2 hours)

(Taken by all candidates.)

17 questions

8 The table shows the approximate risk that women of different ages will have a baby with Down's syndrome.

Mother's age (years)	Approximate risk per 10 000 births
17	4
22	6
27	8
32	11
37	34
42	100
47	217

Source: Dr. J.L. Hamerton

(a) Draw a graph of these figures using the graph paper and the scales overleaf. Join the points that you plot with straight lines. *(4 marks)*

(b) What is the risk that a baby born to a woman of 40 will have Down's syndrome? *(1 mark)*

(c) How many times greater is the risk that a woman will have a Down's-syndrome baby if she is 42 than if she is 22? *(1 mark)*

(d) Suggest two reasons why babies with Down's syndrome are more often born to older mothers. *(2 marks)*

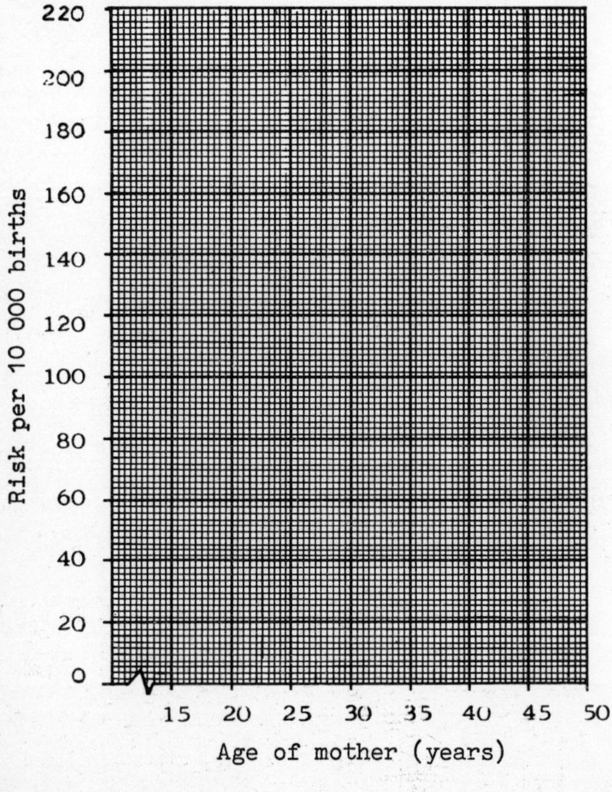

9 A student has been asked to find out how long it takes for starch to be changed to sugar by an enzyme at a temperature of 0°C. The diagram opposite shows the student about to mix known volumes of the enzyme with starch.

(a) What is tube *A* called? *(1 mark)*

(b) Name the substance that will be needed to show that starch has disappeared. *(1 mark)*

(c) Name two other pieces of apparatus that the student should have for this experiment. *(2 marks)*

(d) Suggest two ways the student should improve the part of the experiment shown in the diagram. *(2 marks)*

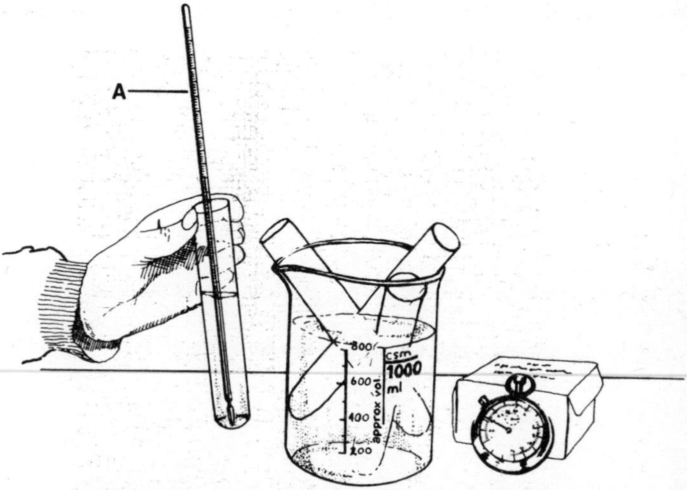

11 The graph below shows the mean body temperatures of two groups of men over a period of four days. Both groups did the same work. One group lived in a cool climate, the other in a hot climate.

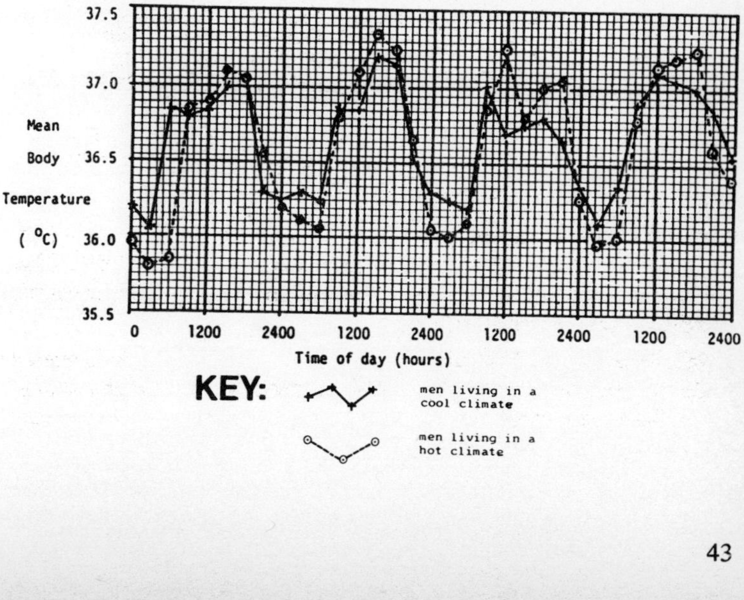

Write down five things that the graph tells you about changes in body temperature. *(5 marks)*

12 The ancestor of the French bean you studied still grows wild in South America. It has a long twining stem, white flowers and a dry pod wall which splits violently to release dry seeds.

(a) Give three ways in which the variety of French bean you studied is different from the wild form described above. *(3 marks)*

(b) Give three processes that plant breeders must have carried out in order to get the variety you studied from the seeds of the wild plant ancestor. *(3 marks)*

14 *(a)* What apparatus is the student in the diagram using? *(1 mark)*

(b) What is this apparatus for? *(1 mark)*

(c) Give an example of what the student might be looking at. *(1 mark)*

(d) What is the student in the diagram doing with her right hand? *(2 marks)*

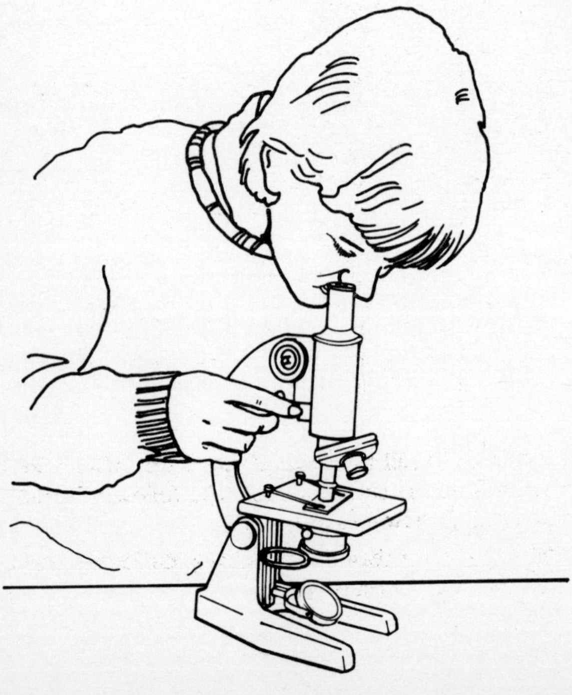

15 What type of habitat did you study?

(a) Name an organism you studied in this habitat.

(i) What apparatus did you use to study where this organism lives? *(1 mark)*

(ii) How should you use this apparatus? *(3 marks)*

(iii) What should you record? *(1 mark)*

(b) Name an environmental factor that you measured in this habitat.

(i) What apparatus did you use to measure this environmental factor? *(1 mark)*

(ii) How should you use this apparatus? *(3 marks)*

(iii) What should you record? *(1 mark)*

17 In late summer, young stems of willow were cut into 30 cm lengths. The lower 10 cm of each stem cutting were stripped of leaves and buds, and then buried firmly out of doors in coarse gritty soil, as shown in the diagram below.

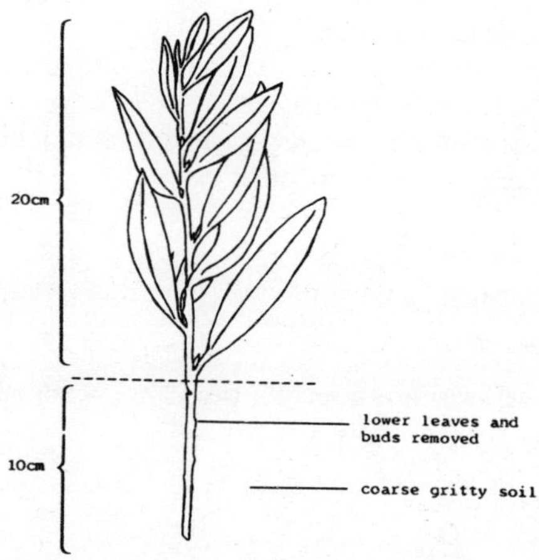

After a few weeks, all the leaves dropped off. After a few months, roots formed on the cut stem ends. The following spring, 70% of the stem cuttings grew into new plants.

(a) *(i)* Suggest one reason for removing the leaves and buds from the lower 10 cm of stem. *(1 mark)* *(ii)* Suggest one reason for using a coarse gritty soil instead of a fine soil. *(1 mark)*

(iii) Suggest one reason for burying each stem firmly (by pressing down the soil around it). *(1 mark)*

(b) *(i)* Suggest one way in which loss of all the leaves after a few weeks might have helped the rooting of the stems. *(2 marks)*

(ii) Suggest one way in which loss of all the leaves after a few weeks might have hindered the rooting of the stems. *(2 marks)*

(c) *(i)* Describe an improvement on this method of growing stem cuttings. *(1 mark)*

(ii) Explain how your improvement would work. *(1 mark)*

(d) A year after the cuttings were put in the soil, one of the new plants was 1.2 m tall with six side branches and another was 0.58 m tall with eleven side branches.

(i) Describe an experiment you could carry out during the following year to help decide whether the differences between these two plants were caused by inheritance or had some other cause. *(4 marks)*

(ii) What results of your experiment would suggest that the differences were inherited? *(1 mark)*

(iii) What results of your experiment would suggest that the differences had some other cause? *(1 mark)*

Paper 3 ($1\frac{1}{2}$ hours)

(Extended level paper.)

6 compulsory questions

1 The diagram below shows a section through a mammal's kidney.

(a) Name the parts *1*, *2* and *3*. *(3 marks)*

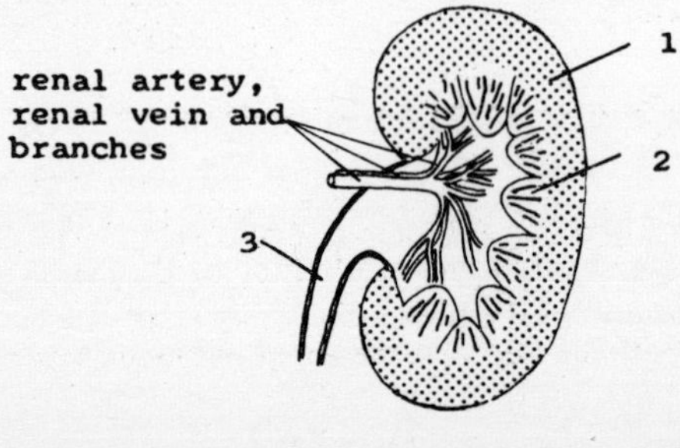

(b) The table below gives information about the human kidney.

(i) What percentage of blood passing into the kidney is filtered into the kidney tubules? *(1 mark)*

(ii) Where in the kidney (part *1*, *2* or *3* on the diagram) does filtration take place? *(1 mark)*

(iii) About 172 dm^3 are filtered from the blood into the kidney tubules per day, yet only 1.5 dm^3 of urine are excreted. What happens to the other 170.5 dm^3? *(2 marks)*

Rate of blood flow in kidneys	Rate of filtration into kidney tubules (nephrons)	Rate of urine passing out of kidneys
1.2 dm^3 per minute	0.12 dm^3 per minute	1.5 dm^3 per day

(c) The table below shows the average amounts of urine, sweat and salt (sodium chloride) lost on a normal day, a cold day and a hot day. (Assume that food and drink are the same on all days.)

(i) Why is more urine lost on a cold day than on a normal day? *(2 marks)*

(ii) Why do you think the total amount of salt lost on each of the three days is the same? *(2 marks)*

(iii) The minimum amount of urine excreted in a day is 0.375 dm^3. Why do you think the kidneys always produce some urine? *(2 marks)*

(iv) What must someone losing more than 7 dm^3 of sweat in a day do in order to remain healthy? *(2 marks)*

	Urine lost per day (dm^3)	Sweat lost per day (dm^3)	Salt (sodium chloride) lost per day	
			in urine (g)	in sweat (g)
Normal day	1.5	0.5	18.0	1.5
Cold day	2.0	0.0	19.5	0.0
Hot day	0.375	2.0	13.5	6.0

2 The map below shows a river flowing through farms *Y* and *Z*.

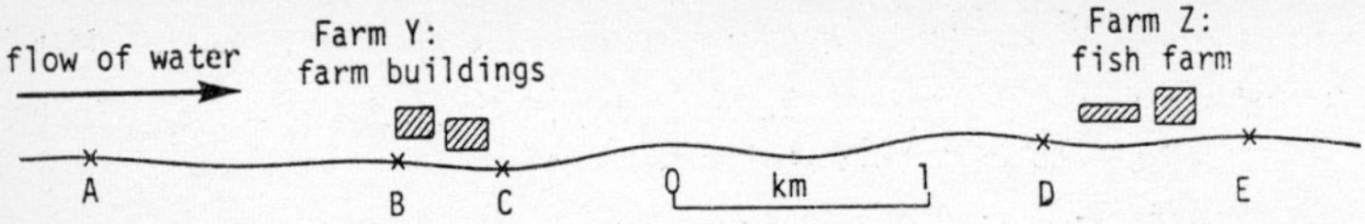

Farmer *Y* changed from beef to dairy cattle. His herd no longer stayed all day in the fields, but came twice a day to the farm buildings to be milked. Farmer *Z* had a fish farm and kept his fish in tanks filled with water from the river. His fish began to die. Scientists analysed the river water at *A*, *B*, *C*, *D* and *E* along the river. Their results are shown in the table.

	River-water analysis		
	Total nitrogen in chemical compounds (parts per million)	pH	Dissolved oxygen (parts per million)
A	0.40	8.5	10.0
B	2.60	6.8	3.6
C	2300.00	4.0	10.0
D	0.76	7.8	1.2
E	0.66	7.8	4.0

(a) *(i)* What three factors shown by the table could have caused the death of Farmer *Z*'s fish? *(1 mark)*

(ii) Which factor do you think the most likely to have caused the death of Farmer *Z*'s fish? Why would this factor have caused the death of his fish? *(2 marks)*

(b) Suggest one explanation each for:

(i) the rise in nitrogen compounds between *A* and *C*; *(2 marks)*

(ii) the fall in nitrogen compounds between *C* and *D*; *(2 marks)*

(iii) the low pH at *C*; *(2 marks)*

(iv) the low oxygen content at *D*. *(2 marks)*

(c) Farmer *Y*'s cows had polluted the river.

(i) What is pollution? *(1 mark)*

(ii) Suggest one way in which Farmer *Y* could reduce the pollution of the river. *(2 marks)*

(iii) Instead of fish, Farmer *Z* wanted to grow water cress in his tanks, but the local health authority would not let him sell it. Why not? *(1 mark)*

3 *(a)* The diagram below shows a section through a mammal's heart and some of its major blood vessels.

(i) What route does blood take from the lungs to the aorta? *(3 marks)*

(ii) Explain how the valve between the left ventricle and the aorta works. *(4 marks)*

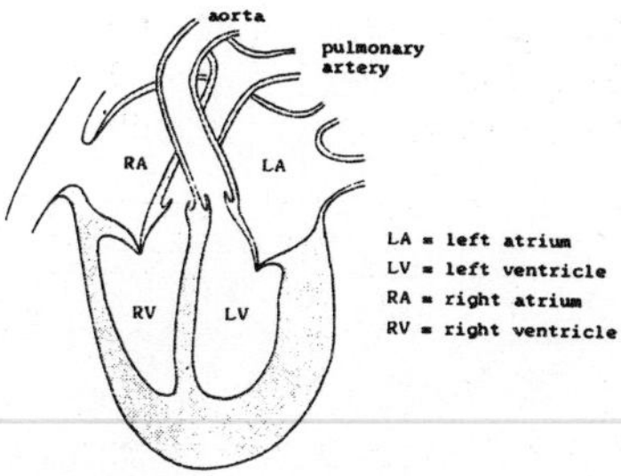

(b) The graph below shows blood pressure in the left ventricle and the aorta throughout two cycles of the heart.

(i) How long is a heart cycle (the time interval between the start of one heartbeat and the start of the next)? *(1 mark)*

(ii) For how long during each heart cycle is the valve between the aorta and the left ventricle closed? How did you work out your answer? *(2 marks)*

(iii) What is the difference between the maximum and minimum pressure in the aorta? *(2 marks)*

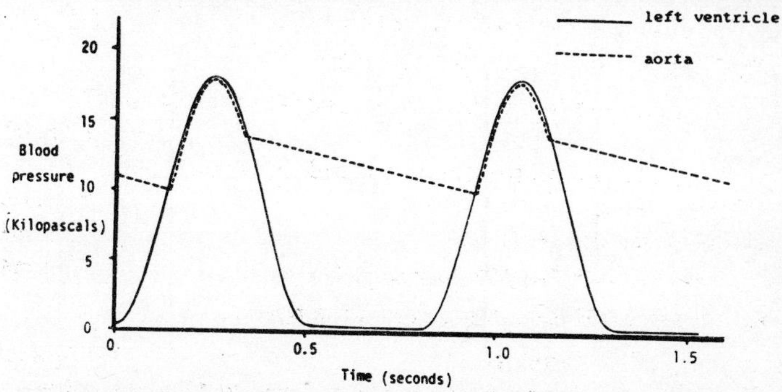

(c) Explain how blood pressure is increased in the left ventricle. *(3 marks)*

LEAG SPECIMEN PAPERS

Paper 1

(Taken by all candidates.)

Alternative A (2 hours)

4 The diagram below shows a young plant growing towards the light, which came from one side only.

(a) In this example:

(i) What was the stimulus? *(1 mark)*

(ii) What was the response? *(1 mark)*

(b) Give *one* advantage to the plant of this response. *(1 mark)*

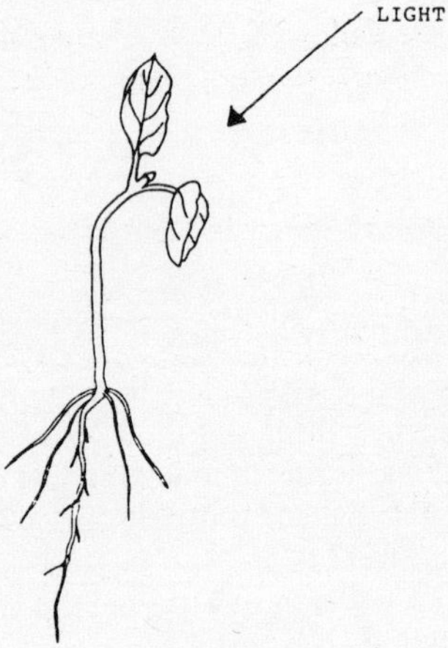

5 The diagram *A* below shows the apparatus for an experiment using dialysis tubing.

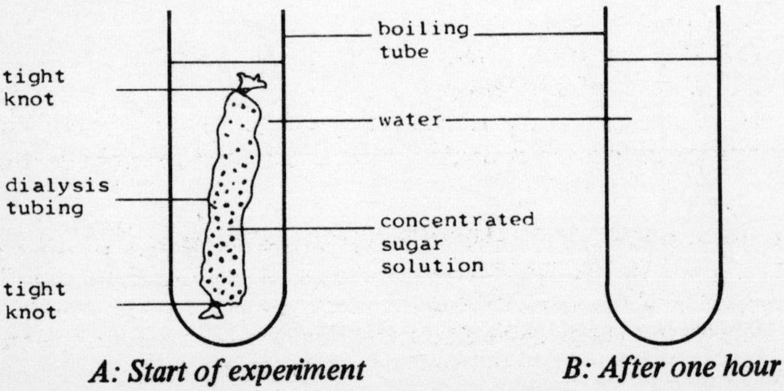

A: Start of experiment *B: After one hour*

Dialysis tubing is flexible and allows very small molecules to pass through it but not larger ones.

(a) *(i)* Complete diagram *B* to show the appearance of the dialysis tubing after one hour. *(1 mark)*

(ii) Explain the change that would occur during the first hour of the experiment. *(2 marks)*

(b) In a second experiment, the same apparatus was set up, but with the water in the dialysis tubing and concentrated sugar solution in the boiling tube. How would both experiments together give you more information about the process which is occurring? *(2 marks)*

9 Starch is a good source of energy.

(a) Explain how the following help to make the starch in food available as an energy source in a mammal:

(i) teeth; *(1 mark)*

(ii) enzymes. *(2 marks)*

(b) Write an equation for the process by which energy is released in cells. *(2 marks)*

10 Which of your sense organs is sensitive to each of the following stimuli?

(a) Music on radio *(1 mark)*

(b) A red traffic light *(1 mark)*

(c) Salt solution *(1 mark)*

12 The kiwifruit originally grew wild in China. In the 1960s a New Zealand farmer planted a number of seeds intending to select plants with the best features.

(a) What name is given to this form of selection? *(1 mark)*

(b) Why was it important for the farmer to use seed from a large number of different plants? *(1 mark)*

(c) The best variety is called Hayward after the farmer who discovered it. Most of the world's cultivated kiwifruit is now the Hayward variety.

(i) How must Hayward plants be reproduced to keep the variety the same? *(1 mark)*

(ii) What could research workers do if they want to improve on the Hayward variety? *(2 marks)*

Alternative B (2 hours)

1 The diagram below shows the layers in a pond.

(a) Name the layers shown by the letters *A*, *B* and *C*. *(3 marks)*

(b) *(i)* Name an example of a plant you would expect to find in layer *B*. *(1 mark)*

(ii) State how the light intensity in layer *B* differs from the light intensity in layer *A* during daylight. *(1 mark)*

(iii) How might this difference in light intensity affect the plant you have named? *(2 marks)*

Paper 3A (1 hour)

(Designed for candidates expected to gain grades A to D.)

1 A 22-year-old woman kept a precise record of her food and drink intake for one day. The quantities of some major nutrients were calculated, and the totals compared with the average daily requirement of a woman of that age. The figures are summarised in the table below.

Meal	Item	Quantity	Energy in kJ	Protein in g	Fat in g	Carbohydrate in g	Calcium in mg	Iron in mg	Vitamin C in mg
Breakfast	white bread	90 g	950	7	2	50	90	1	0
	butter	15 g	450	0	12	0	2	0	0
	jam	30 g	330	0	0	19	5	0	1
	black coffee	1 cup	20	0	0	1	4	0	0
Lunch	hamburger	150 g	1560	30	15	30	50	4	0
	ice-cream	100 g	800	4	12	20	130	0	1
	fizzy drink	1 can	550	0	0	30	0	0	0
Evening meal	sausages	75 g	1150	9	24	10	30	0.5	0
	chips	200 g	2100	8	20	70	25	2	20
	baked beans	220 g	600	10	1	20	100	2.5	4
	apple pie	150 g	1800	5	25	60	60	1	1
	cream	30 g	550	0.5	15	1	20	0	0
	tea with milk	2 cups	200	2	4	6	100	0	0
Snacks	chocolate	50 g	1200	5	20	25	120	1	0
	peanuts	50 g	1200	15	25	5	30	1	0
TOTAL INTAKE FOR DAY			13460	95.5	175	347	766	13	27
AVERAGE DAILY REQUIREMENT			9400	58	*	*	600	14	30

*Amounts variable

(a) *(i)* By how much did the energy content of the day's diet exceed the average daily energy requirement of a 22-year-old woman? *(1 mark)*

(ii) What would be the probable effect of this difference on an average woman of this age over a long time? *(1 mark)*

(iii) This woman could have a daily energy requirement much greater than average. Suggest *two* reasons why this would be so. *(2 marks)*

(b) The recommended carbohydrate : fat ratio in a balanced diet is 5:1 by weight.

(i) Which individual *meal* in the day given in the table had a carbohydrate : fat ratio of exactly 5:1? *(1 mark)*

(ii) To the nearest whole number, what is the carbohydrate : fat ratio for the whole day's intake? *(1 mark)*

(iii) What effects may this proportion of fat in the daily diet have on the woman's circulatory system? *(2 marks)*

(c) *Excluding* coffee, tea and fizzy drinks, which of the items eaten during the day had the highest level of calcium per gram of food? Give the calcium content of this item in mg per g of food. *(2 marks)*

(d) Identify *one* nutrient in which the day's food intake is *deficient* and name a deficiency disease that may result if the woman's diet continues to provide too little of the nutrient. *(1 mark)*

(e) A sample of the woman's urine taken the morning *after* the day described above contained a high concentration of urea. How might this have been predicted from the information in the table? *(1 mark)*

MEG SPECIMEN PAPERS

Paper 2 (1 hour 20 minutes)

(Taken by all candidates.)

Answer all questions.

Section A

8 questions

6 The figures in the table overleaf show the composition of air breathed in and out in man.

(a) *(i)* Which one of the gases is used for respiration? *(1 mark)*

	inspired air/%	expired air/%
gas A	20.6	16.6
gas B	0.04	4.0
gas C	79.0	79.0

(ii) Name gas *B*. *(1 mark)*

The composition of air in the atmosphere remains the same, yet the table above indicates that the composition of the air should differ as a result of breathing in and out. How is the balance in the atmosphere maintained? *(2 marks)*

(iii) Why is the water vapour content greater in expired than in inspired air? *(1 mark)*

(b) State *one* feature of the lungs which allows the rapid exchange of gases. *(1 mark)*

(c) State *one* feature of leaves which allows the rapid exchange of gases. *(1 mark)*

(d) Name *one* substance found in cigarette smoke that may harm the gaseous exchange surface in the lungs. *(1 mark)*

(e) Under what conditions is lactic acid produced in muscle tissue? *(2 marks)*

Section B

7 questions

15 The diagrams which follow illustrate population by age structure in a South American country and in the British Isles at the present day.

(a) Suggest a reason for:

(i) the relatively high birth-rate in the South American country; *(1 mark)*

(ii) the relatively high death-rate in the South American country. *(1 mark)*

(b) The death-rate amongst reproductive adults in the British Isles is now much lower than it was in 1850. Suggest *two* reasons for this decrease. *(2 marks)*

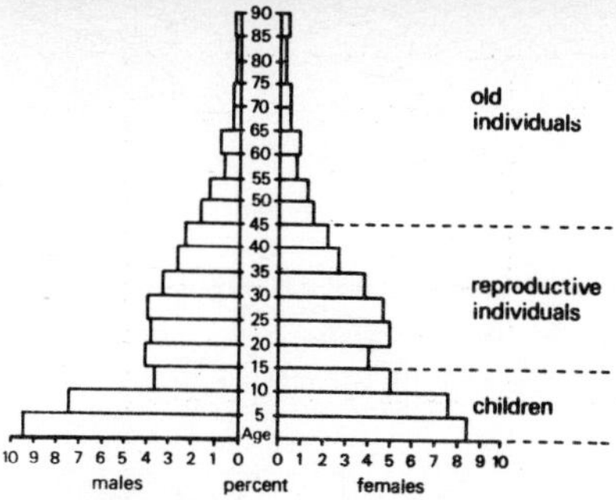

A South American country

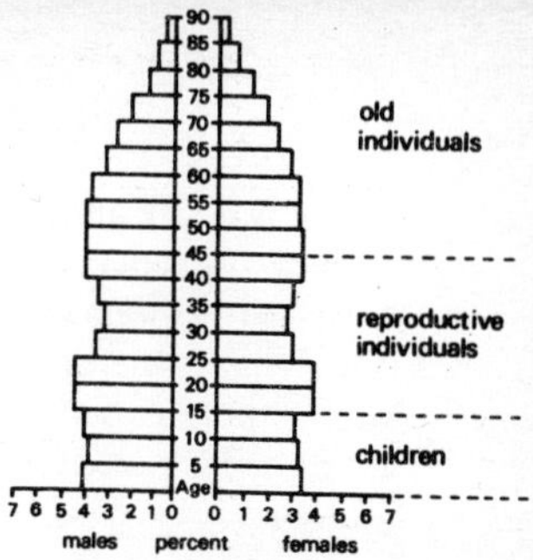

The British Isles

(c) State two ways in which the population of old individuals in the South American country differs from that in the British Isles. *(2 marks)*

Paper 3 (1 hour 15 minutes)

(Optional paper for candidates expected to gain grades A or B.)

Section A

7 questions

Answer all questions.

1 An experiment was performed to investigate the effect of light on the growth of the coleoptiles (shoot tips) of oat seedlings.

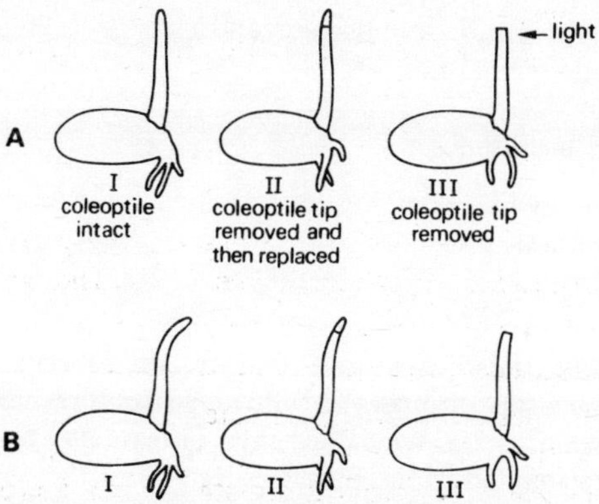

In the diagram, row *A* shows three seedlings at the beginning of the experiment. The seedlings were exposed to light from one side for three hours. Row *B* shows the results.

(a) What is the name given to the response shown by seedlings I and II? *(1 mark)*

(b) How does this experiment support the idea that the response is produced by auxins (plant hormones)? *(2 marks)*

(c) State *two* properties possessed by auxins. *(2 marks)*

(d) A similar response is shown by leafy shoots to light from one side. What is the importance of this response? *(2 marks)*

(e) What stimulus causes roots to respond to auxins if the roots are placed horizontally? *(1 mark)*

Section B

4 questions

Answer any two of the following questions.

4 ***(a)*** The liver and the pancreas have important roles to play in the digestion of food. Describe these roles in detail. *(10 marks)*

(b) The liver and the pancreas also have roles beyond that of digestion.

Describe any ways in which these organs are involved in:

(i) the storage of carbohydrates; *(4 marks)*

(ii) the excretion of wastes. *(6 marks)*

5 'Man has brought more and more biological knowledge to bear upon the problem of producing more food, and food of the right quality to keep him healthy.'

Discuss the role of the following, in this context:

(a) protein production by micro-organisms;

(b) butter and margarine;

(c) legumes;

(d) artificial fertilisers. *(20 marks)*

6 ***(a)*** Compare, by means of concise statements, the following terms:

(i) diffusion;

(ii) osmosis;

(iii) active uptake. *(9 marks)*

(b) It is biologically important that certain substances enter or leave organisms through the following structures. Explain what the substances are, why they enter or leave and how the structures are adapted to the purposes they fulfil:

(i) root hairs;

(ii) leaves;

(iii) villi. *(11 marks)*

WJEC SPECIMEN PAPER GCE/CSE

Paper 1 (2½ hours)

(Designed for candidates expected to gain grades G to C.)

22 questions

Answer all questions.

9 The first diagram below shows the nucleus of an animal cell at the start of reduction division (meiosis).

(a) Name an organ where meiosis takes place.
(b) Name the structures shown inside the nucleus.
(c) Complete the second diagram to show a nucleus produced after the reduction division. *(3 marks)*

15 *(a)* Label the parts indicated on the two half-flowers shown in the diagrams which follow.
(b) State the function of the structure labelled *3*.
(c) State *one* function of the structure labelled *8*.
(d) State *two* differences, visible in the diagrams, between flowers *A* and *B*.
(e) How many stamens does flower *B* possess? *(10 marks)*

A

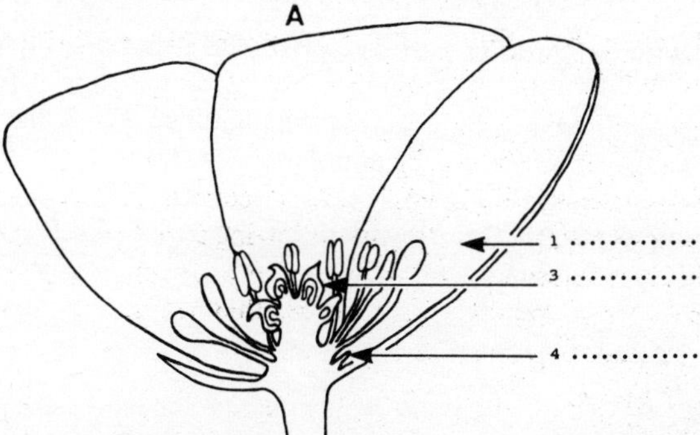

B

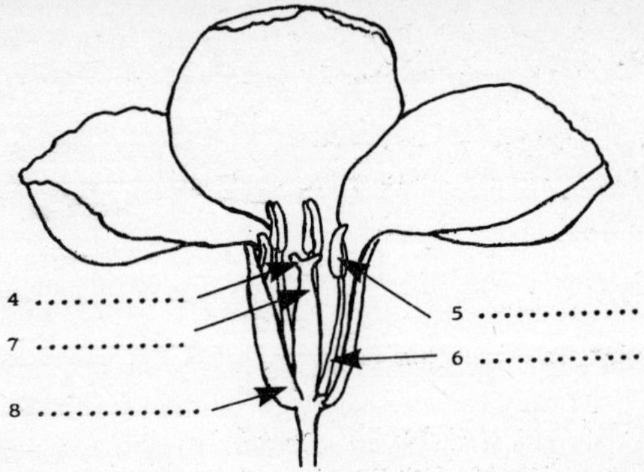

17 The table shows the estimated population of the world since the year AD 1.

Year (A.D.)	Population (millions)
1	200
1000	300
1500	400
1800	1000
1900	2000
1950	3000

(a) Use these figures to plot a graph on the paper provided. *(4 marks)*

(b) Suggest *two* possible factors which, since 1800, have caused the change in the rate of population increase. *(2 marks)*

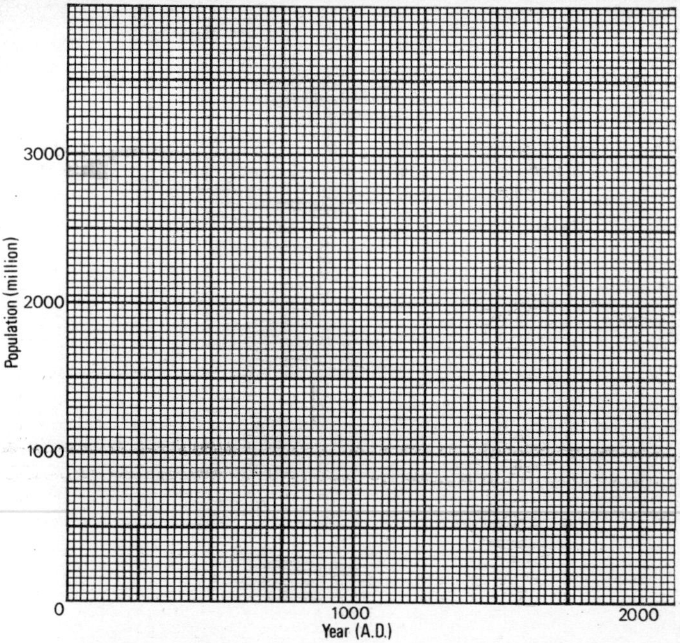

18 Each year in Britain, 250 tonnes of lead from fishermen's weights reach the environment. It is feared that some types of bird suffer from this form of pollution because they swallow the lead weights during feeding. Study the data obtained between 1973–80 for mute swans, given below, and answer the questions which follow. *Note*: 40 g/100 cm^3 is considered relatively harmless in swans.

Average concentration of lead in blood of swans g/100cm^3	Area inhabited
106.8	Trent (near a town centre)
28.8	Dorset (countryside)
104.6	Thames (near town centre)
107.4	Avon (near town centre)
18.6	Mid Wales (countryside)
16.4	Tweed (countryside)

(a) Using *only* the data given, suggest *one* reason why there is more lead in the blood of swans from near towns than from swans in the countryside.

(b) Explain how the lead reached the swans' blood.

(c) Lead affects the nervous tissues and the muscle tissues of vertebrates. Symptoms of lead poisoning in swans include blindness and the oesophagus packed with food.

(i) Name the layer of the eye which you would expect to be affected.

(ii) Suggest *two* reasons why food is found packed in the oesophagus.

(d) Suggest *two* ways of reducing this pollution without banning fishing *altogether*. *(7 marks)*

19 During a study of habitat, pupils collected some molluscs. These are drawn below. Underneath the drawings is an identification key.

(a) Use the key to identify *B*, *D* and *F*.

(b) Use the key to describe *three* features of *Planorbis*. *(6 marks)*

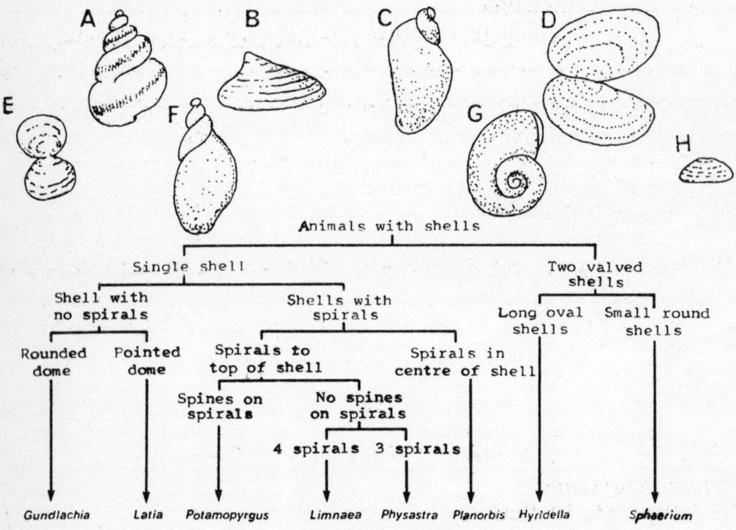

21 (a) Complete the diagram which follows by drawing the pathway of the reflex arc in the following stages.

(i) Draw single lines to show the following nerve fibres: sensory (afferent); relay (intermediate); motor (efferent).

(ii) *Label* the nerve fibres clearly.

(iii) Draw arrows on the nerve fibres to show the direction of a nerve impulse.

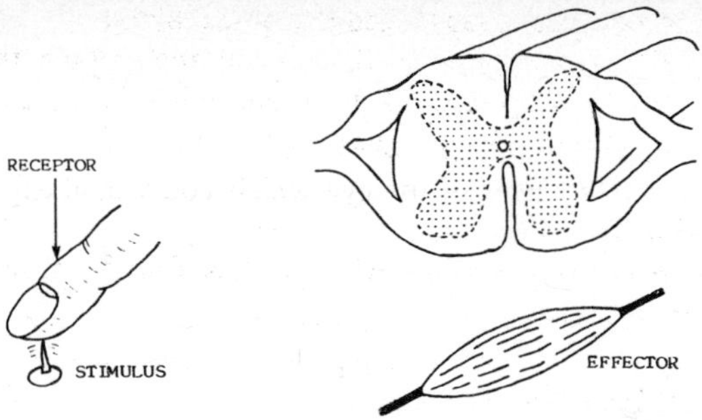

(b) List *three* differences between a reflex action and an action brought about by hormones.

(c) When the impulse reaches the effector, it causes it to contract.

(i) Name *two* substances required to provide energy for contraction.

(ii) State how these substances are carried to the effector.

(11 marks)

Paper 2 is designed for candidates expected to gain grades A to C. A specimen question and sample answers are given in section 5.

INDEX TO QUESTIONS

The following list gives page references for questions on specific parts of the syllabus.

Diversity of organisms

Classification and keys
NEA Paper IB, question 6; WJEC Paper 1, 19 — 31; 60

Relationships between organisms and with the environment

Natural ecosystems
NEA Paper IA, 10; SEG Paper 1, 15; LEAG Paper 1B, 1 — 29; 45; 52
Human influences: conservation; pollution
NEA Paper IIP, 2; Paper IIQ, 2 and 3; SEG Paper 3, 2; WJEC Paper 1, 18 — 34; 36; 48; 59
World human populations
NEA Paper IA, 14; MEG Paper 2B, 15; WJEC Paper 1, 17 — 29; 54; 58

Organisation and maintenance of the individual

Cell structure and organisation	
SEG Paper 1, 14	44
Chemistry of life	
NEA Paper IB, 9; SEG Paper 1, 9; LEAG Paper 1A, 5	32; 42; 50
Plant nutrition	
NEA Paper IA, 25–8; Paper IIQ, 6	30; 38
Animal nutrition	
NEA Paper IB, 12; LEAG Paper 1A, 9; Paper 3A, 1; MEG Paper 3B, 4 and 5	33; 51; 52; 56
Transport in plants	
SEG Paper 1, 17; MEG Paper 3B, 6	45; 56
Transport in animals	
SEG Paper 3, 3	49
Energy release	
NEA Paper IB, 14; MEG Paper 2A, 6	33; 53
Homeostasis: skin, liver, kidney	
NEA Paper IIP, 7; SEG Paper 1, 11; Paper 3, 1	35; 43; 46
Support and movement	
NEA Paper IIP, 8	35
Receptors and nervous coordination	
NEA Paper IIQ, 10; LEAG Paper 1A, 10; WJEC Paper 1, 21	38; 51; 60
Hormonal control	
LEAG Paper 1A, 4; MEG Paper 3A, 1	50; 55

Development of organisms and the continuity of life

Reproduction in plants	
NEA Paper IA, 15; WJEC Paper 1, 15	29; 57
Reproduction in animals	
NEA Paper IIQ, 12	39
Growth and cell division	
NEA Paper IA, 16; WJEC Paper 1, 9	30; 57
Variation and inheritance	
NEA Paper IA, 17; Paper IIP, 12; Paper IIQ, 13; SEG Paper 1, 8 and 17; LEAG Paper 1A, 12;	30; 36; 40; 41; 45; 51
Selection	
NEA Paper IB, 7; Paper IIQ, 15; SEG Paper 1, 12	31; 40; 44

SECTION 5

Specimen answers

There are no 'right' answers to many exam questions, as they can be answered at different levels. It won't help you to learn one model answer for the exam, because you need to understand exactly what the examiners are looking for.

Your answer will be judged by the degree of:

(a) knowledge and understanding;
(b) skills and scientific processes

that you have shown. The exam groups have worked out exactly what level of each they would expect to see in order to award a particular grade. For example, here is a sample question followed by specimen answers at each grade.

SEG SPECIMEN PAPER

Paper 3

6 **Two kilograms of tomatoes were bought at the same time from the same batch in the same shop. One kilogram was kept in an air-tight plastic (polythene) bag for a week, and four of these tomatoes 'went bad'. The other kilogram was left open to the air, and none of these tomatoes 'went bad'.**

(a) **From your knowledge of the process of decomposition, describe what happens when a tomato 'goes bad'.** ***(5 marks)***

Answers

Grade A: In moist conditions, the fungal spores on the surface of the tomato enter through the damaged skin and germinate into the mycelium, which grows by secreting enzymes into the flesh of the tomato to break down the fruit tissues. The breakdown products give the bad smell and taste to the tomato.

Grade B: In moist conditions, fungi on the surface enter the damaged tomato skin and break down the tomato tissue.

Grade C: In moist conditions, fungi grow on the tomato skin. The mycelium breaks down the tomato.

Grade D: Fungi grow on the skin and break down the tomato.
Grade E: Fungi grow on the tomato.

Compare these carefully, and work out for yourself how the answers differ.

On the following pages you will find more graded answers; try answering the questions yourself first, then compare your answer with the different grades so that you can estimate what level you might have achieved so far. Unless the question says otherwise, your answers should be written in full sentences. Where a detailed explanation is required, you will need to plan your answer carefully to cover the relevant points concisely and coherently: notes are not sufficient.

(b) **Suggest two hypotheses to account for the fact that some tomatoes 'went bad' while others in the same batch did not.** ***(2 marks)***

Answers

Grade A: Any two valid hypotheses correctly stated. For example: 'Damaged tomatoes in the batch went bad and undamaged ones did not,' or 'The tomatoes touching the polythene went bad but the others did not.'
Grade B: Two hypotheses but incomplete, e.g. omitting 'undamaged ones did not'.
Grade C: Any one hypothesis correctly stated.
Grade D: Any one hypothesis but incomplete.

(c) **Choose one of your hypotheses and describe how you would carry out an experiment to test it. Give details of the apparatus you would use and of the conditions in which you would conduct your experiment. What would you measure and how would you record your results?** ***(8 marks)***

Answers

Grade A: To test the hypothesis: 'Tomatoes in airtight conditions go bad, but the others do not.'
Apparatus: 2 sterile plastic bags; 1 metal tie.
Method: Take three batches of the same variety of tomatoes with ten per batch. Pick the tomatoes at the same time and of the same size. Check they are undamaged, with no mould visible. Place one batch of ten in one plastic bag and seal it with the metal tie. Place another batch in an open bag, and place the last batch on the bench. These two batches are controls. Leave the three batches at the same temperature and light intensity on the laboratory bench for one week. Examine them each day. Note the number with fungi growing, in the following table:

Time in days	**Number of tomatoes going bad** *Open bag*	*Closed bag*	*No bag*
1			
2			
3			
etc.			

Grade B: As for grade A, but omitting precautions such as 'picking tomatoes of the same variety at the same time and of the same size'; 'two controls'; 'sterile bags'; 'checking they are undamaged with no mould visible' and 'Leave the batches at the same temperature and light intensity on the laboratory bench.'
Grade C: Put some tomatoes in a plastic bag and seal with a metal tie. Put some more tomatoes in another bag and leave it open. Leave for one week. Note the number going bad in each bag.
Grade D: Put some tomatoes in a plastic bag and seal with a metal tie. Leave some more tomatoes on the bench. Leave for one week. Note the number going bad in each batch.
Grade E: As in grade D, but not leaving for a suitable time.

5 **Oestrogen and progesterone are two female sex hormones. The first diagram shows some of the events in a 28-day menstrual cycle: the width of the outer band indicates the relative quantity of sex hormones in the blood.**

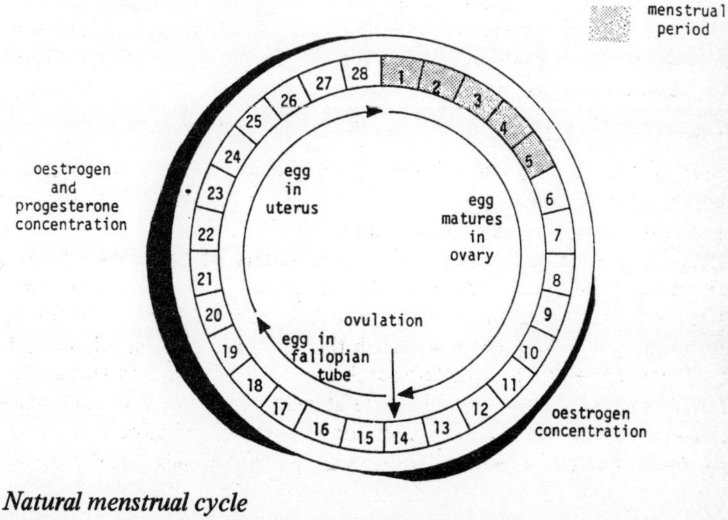

Natural menstrual cycle

(a) (i) **On which three days of this cycle would fertilisation of the egg be most likely?** ***(1 mark)***

Answer: 14 to 16, or 15 to 17.

(ii) **If sperms can live in the female reproductive organs for up to eight days, on which days could insemination lead to fertilisation?** ***(1 mark)***

Answer: 7 to 16, or 8 to 17 days.

(b) **How does the concentration of sex hormones appear to control events in the menstrual cycle?** ***(3 marks)***

Answer: High oestrogen level falls before ovulation; oestrogen and progesterone levels both rise before egg reaches the uterus; concentration of both fall before menstruation; or any three statements valid from diagram. Each valid statement scores 1 mark.

(c) **Substances like oestrogen and progesterone are now factory-made and are sometimes used to regulate a menstrual cycle. The second diagram shows one form of this treatment (though the relative quantities of sex hormones in the blood are not shown).**

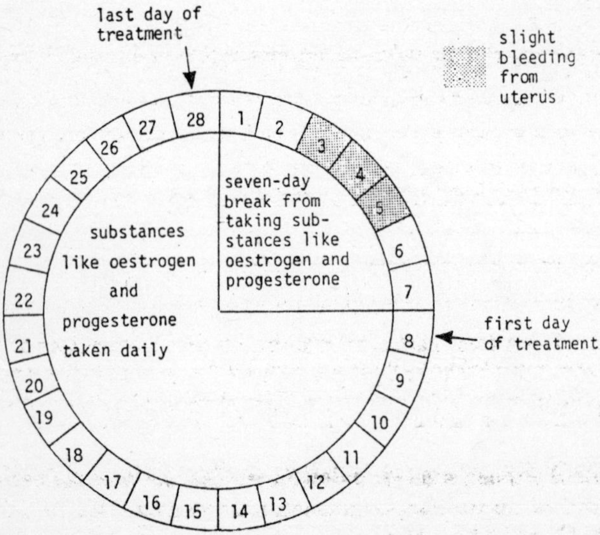

Treatment cycle

Compare events in the natural 28-day menstrual cycle and the 28-day treatment cycle and describe similarities and differences wherever you can. ***(10 marks)***

Answer:
Similarities are: both 28-day cycles; bleeding in both; oestrogen and progesterone in both.
Differences are: bleeding less in pill cycle; shorter time of bleeding in pill cycle; no egg present in pill cycle; hormone levels renewed daily in pill cycle. *1 mark each = 7 marks*
Possible reasons for the differences are: fall in hormones results in bleeding; less bleeding because no egg is present; progesterone stops the egg from developing. *1 mark each = 3 marks*

Grade A: a total of 12 or more marks.
Grade B: 9–11 marks.
Grade C: 7–8 marks
Grade D: 5–6 marks.
Grade E: 4 marks.
Grade F: 3 marks.

LEAG SPECIMEN PAPERS

Paper 1

15 **The diagram shows the external appearance of a mung-bean seed.**

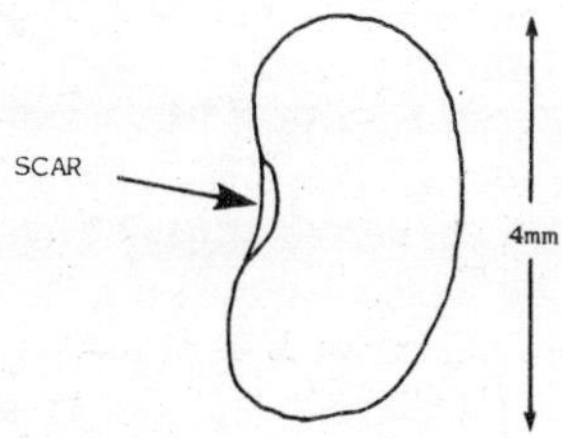

(a) **From what part of the flower has the seed developed?** ***(1 mark)***

Answer: Ovule.

(b) **To what part of the flower was the seed attached at the scar?** ***(1 mark)***

Answer: Ovary or fruit.

(c) **You have been asked to find the best temperature for sprouting (germinating) mung-bean seeds. You have been given 100**

seeds, ten test-tubes, cotton-wool, ten thermometers, a clock and a supply of water.

(i) **Draw and label one of the test-tubes you would set up for the experiment.** ***(2 marks)***

Answer: At least five seeds per test-tube; seeds on moist cotton-wool or suitable provision for moisture.

(ii) **Explain how you would carry out this experiment.** ***(7 marks)***

Answers

Grade A: Soak the seeds and divide into ten (or three) batches of ten (or five). Place each batch on wet cotton-wool in a test-tube. Keep each test-tube at a different but constant temperature, at 5°C intervals from 5°C to 50°C (or at 20°C intervals if only three batches, e.g. 5°C; 25°C and 45°C). Add ice to the water bath at low temperatures and warm water at higher temperatures, or use electric water-baths with thermostats. Keep other conditions constant, e.g. light intensity; aeration. Note the time. Examine every twelve hours for one week. Record the number germinated by noting the number of seeds with a split testa or a radicle appearing.

Grade B: Divide the seeds into ten batches (or three) of ten (or five). Place each batch on wet cotton-wool in a test-tube. Keep each test-tube at a different temperature at 5°C intervals from 5°C (if three batches: 5°C; 25°C and 45°C). Note the number of seeds germinated at each temperature after five days.

Grade C: Divide seeds into ten (or three) batches of ten (or five). Place each batch on wet cotton-wool. Keep the tubes at different temperatures. Note the number of seeds germinating in each tube.

Grade D: Place the seeds on wet cotton-wool in test-tubes at different temperatures. Note the number of seeds germinating in each tube.

Grade F: Keep the seeds at different temperatures. Note the number of seeds germinating at each temperature.

(iii) **Describe one difficulty you might have in carrying out this experiment.** ***(1 mark)***

Answer: Keeping the temperature constant; examining often enough; or any sensible suggestion. *Any one suggestion = 1 mark*

Paper 2

(Maximum grade C.)

3A **Look at the diagram.**

(a) **Name the parts** ***A*****,** ***B*** **and** ***C*****.** ***(3 marks)***

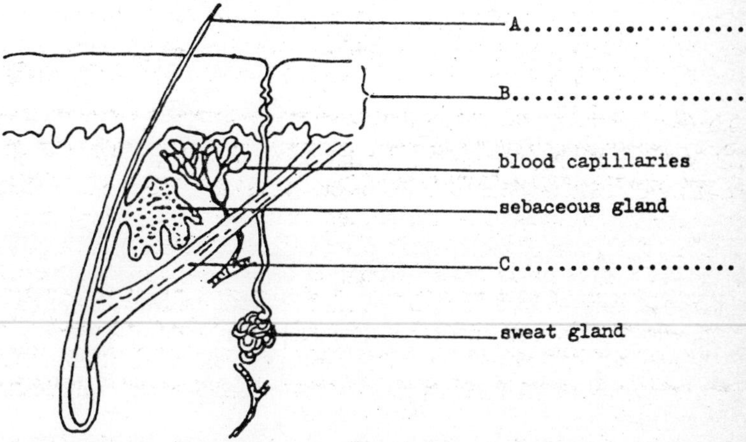

Answer: A = hair; *B* = epidermis; *C* = erector muscle.

(b) **After you have run a race, your skin helps you cool down.**

(i) **Explain how changes in the blood capillaries in the skin help to cool your blood.** ***(4 marks)***

Answers

Grade C: Blood vessels near the surface of the skin dilate, so that more blood flows near the surface and loses heat by radiation or conduction.
Grade F: More blood flows near the surface of the skin.

(ii) **Explain how the sweat glands help to cool your blood.** ***(4 marks)***

Answers

Grade C: Sweat solution is secreted from the sweat gland along the duct to the surface of the skin. The water evaporates absorbing the latent heat of vaporisation from the skin and cooling the skin.
Grade F: Sweat evaporates.

(c) **How do you think the sebaceous glands help to keep hair and skin healthy?** ***(3 marks)***

Answers

Grade C: The sebaceous gland secretes grease which keeps the skin supple, preventing cracking and infections. The grease reduces evaporation from the skin and makes water run off the hairs.
Grade F: Sebaceous glands secrete grease.

3B **Look at the diagram.**

(a) **Name the parts** ***A***, ***B*** **and** ***C.*** ***(3 marks)***

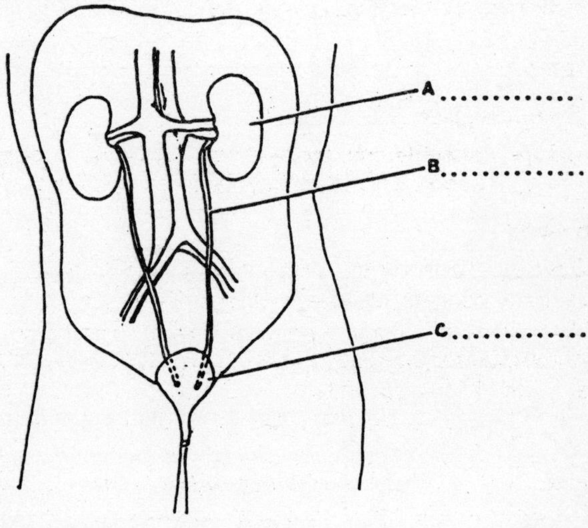

Answer: A = kidney; *B* = ureter; *C* = bladder.

(b) *(i)* **Where is urea formed?** ***(1 mark)***

Answer: Liver.

(ii) **Where and how is urea separated from the blood?** ***(3 marks)***

Answers

Grade C: A solution of urea is filtered off into the capsule of the nephron from the glomerulus by ultrafiltration.
Grade F: The urea is filtered off.

(c) **Explain why less urine is passed out on a hot day than on a cold day.** *(4 marks)*

Answers

Grade C: More water is lost in the sweat on a hot day to cool the body. The kidney regulates the water balance of the body by losing less water in the urine.
Grade F: More sweating on a hot day.

(d) **When a person has a diseased kidney removed, what *three* parts must be cut and sealed off?** *(3 marks)*

Answer: renal artery; renal vein; ureter.

Paper 3

(Maximum grade A.)

2A (a) **Describe one method you could use to investigate the distribution of a named animal in a deciduous wood.** *(5 marks)*

Answers

Grade A: Earthworms. Select a number of different areas in a woodland; for example, areas of high and low light intensity at ground level. Place ten 1 m^2 quadrats in each region. Spray one gallon of 0.2% potassium permanganate solution into each quadrat using a watering-can. Wait 30 minutes. Place the earthworms in a jar as they emerge and count them. Replace them. Compare the total number found in each area.
Grade B: Place 1 m^2 quadrats in different areas in a woodland. Spray one gallon of dilute potassium permanganate solution into each quadrat. Wait 30 minutes. Count the earthworms emerging. Compare the total number found in each area.
Grade C: Place 1 m^2 quadrats in different areas in a woodland. Pour dilute potassium permanganate solution into each quadrat. Count the number of earthworms emerging.
Grade D: Pour dilute potassium permanganate solution into 1 m^2 quadrats. Count the number of earthworms emerging.
Grade F: Count the number of earthworms in quadrats.

(b) **Explain the influence that humans and other animals living in the area surrounding a wood might have on the woodland ecosystem.** *(5 marks)*

Answer

Humans fell or coppice trees. They collect firewood and pick flowers. They trample paths and may shoot birds. They spray crops with pesticides which may blow into the woodland, killing insects. Owls may prey on woodmice. Bees pollinate flowers. Herbivores, such as rabbits, graze on grass. The faeces of these animals provide food for fungi. Birds may disperse berries, such as holly.

Grade A: Five examples; at least one from other animals and one from humans.
Grade B: Four examples; at least one from other animals and one from humans.
Grade C: Three examples; at least one from other animals and one from humans.
Grade D: Two examples; at least one from other animals and one from humans.
Grade E: Any two examples.
Grade F: Any one example.

(c) **Explain how acid rain is caused and what effects it might have on woodland. Suggest ways in which acid rain could be reduced.** ***(4 marks)***

Answers

Grade A: Carbon dioxide and oxides of sulphur and nitrogen from combustion dissolve in rainwater making an acidic solution. This acid rain kills lichens and decreases the pH of soil and leaf litter. Micro-organisms are less active at low pH, so the rate of decomposition falls.

Any 3 points score maximum 3 marks.

Control by less combustion or by legislating smokeless zones, filters on chimneys or car exhausts.

Any 2 points score maximum 2 marks.
But maximum total for ***(c)*** *= 4 marks*

Grade B: Three examples; at least one from each part.
Grade C: Two examples; at least one from each part.
Grade D: Any two examples.
Grade F: One example.

2B (a) **Describe how energy enters, passes through and is lost from a woodland ecosystem.** ***(10 marks)***

Answers

Grade A: Light from the sun is absorbed by the chlorophyll in green plants. In the plant chloroplasts, the light energy is converted to chemical energy by photosynthesis. This chemical energy is then used to build up carbon dioxide and water into carbohydrates, such as starch.

When the plants are eaten by herbivores, such as aphids, the chemical energy is transferred to animals. Herbivores are eaten by carnivores, such as ladybirds, or destroyed by decomposers, such as fungi, so the energy is passed into other animals or plants.

Some chemical energy is passed into the soil as urine or faeces by consumers. The faeces and dead bodies are decomposed by fungi or micro-organisms.

Some energy is lost from an ecosystem by all the organisms, including micro-organisms, as heat from respiration. Energy is also lost when the animals migrate from the area or are preyed upon by animals from another area.

Grade C: Light energy is absorbed by plants and built up into carbohydrates in photosynthesis.

The plants are eaten by herbivores which are eaten by carnivores. Energy is lost in respiration as heat.

Grade E: Light is absorbed by plants in photosynthesis. The plants are eaten by animals which lose energy in respiration.

(b) **Why are woodland ecosystems being lost? What measures can be taken to conserve them?** ***(4 marks)***

Answer

Land is needed for agriculture and building. Wood is also needed for building and for wood products, such as paper. Acid rain slows down decomposition. Woodlands may be badly managed, for example by not replanting trees. The use of pesticides and the irresponsible dumping of waste reduce the numbers of some animals and plants.

Woodlands can be conserved by governments' giving money to replant trees. Laws can be passed to prevent trees being felled. Nature reserves may be set up. Saplings can be protected from grazing by fences.

Grade A: Four points; at least one from each section.
Grade B: Three points.
Grade C: Two points
Grade D: One point.

WJEC SPECIMEN PAPERS GCE/CSE

Paper 1

(Maximum grade C.)

20 *(a)* **Complete the table below to show how each of the diseases are spread.**

Write *A* if spread by *air*.
Write *I* if spread by *insects*.
Write *S* if spread by *sexual contact*.
Write *N* if the disease is not spread by *any of these methods*.

Disease	***Method of spread: A, I, S or N***
malaria	
anaemia	
tuberculosis	
athlete's foot	
venereal disease	
influenza	

(3 marks)

Answer: malaria: *I*; anaemia: *N*; tuberculosis: *A*; athlete's foot: *N*; venereal disease: *S*; influenza: *A*. $\frac{1}{2}$ *each = 3 marks*

(b) **Which of the diseases listed in the table is caused by a virus?** ***(1 mark)***

Answer: influenza.

(c) **Describe the main methods used in the control of malaria.** ***(6 marks)***

Answer:
1 Spray mosquitoes with insecticides.
2 Drain swamps.
3 Remove upturned cans which collect water.
4 Spray swamps with oil.
5 Remove vegetation from around villages.
6 Sterilise male mosquitoes with ultraviolet light and release them.
7 Take drugs, such as mepacrine, daily to kill the malarial parasite.
8 Cover windows and beds with mosquito nets.

Any 6 score 1 each to 6 marks.

Grade C: a total of 7 to 10 marks.
Grade D: 5 to 6 marks.

Grade E: 4 marks.
Grade F: 3 marks.

Paper 2

(Maximum grade A.)

17 *(a)* **Make *two* labelled diagrams of transverse sections to show the typical structure of a herbaceous stem and root.**

(8 marks)

Answers

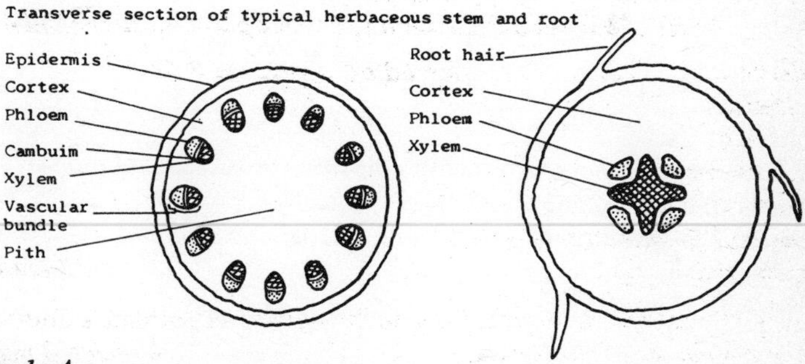

Grade A

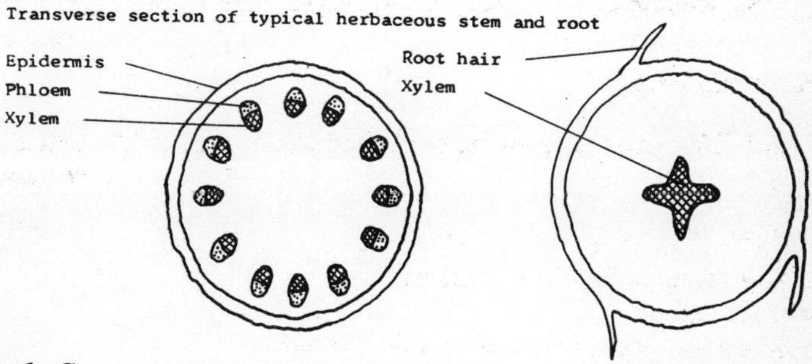

Grade C

(b) **Explain how the living tissues help to support the stem.**

(4 marks)

Answers

Grade A: Inner cells are turgid under a compression strain. Outer epidermal cells form a continuous layer under an extension strain. These equal, but opposite, forces give a rigid structure.

Grade C: Cells are turgid and compressed to give a rigid structure.

(c) **Describe under suitable headings a single experiment to demonstrate osmosis in a living system.** *(6 marks)*

Answers

Grade A

Prepare tissue: Strip the epidermis from rhubarb.
Treat tissue: Take two pieces of epidermis and immerse one in a watch-glass of molar sucrose and one in pure water. Leave for 15 minutes.
Examination: Mount the epidermis in the solution it was immersed in. Add a cover slip and examine it under high power of the microscope.
Result: Note the cell contents have contracted from the wall in the sucrose solution, but not in the pure water.

Grade C

Method: Place a piece of rhubarb epidermis in sucrose solution and a piece in water. Examine these under the microscope.
Result: The contents of the cells in sucrose solution had contracted from the wall.

SECTION 6

Assessment and marking

ASSESSMENT

The previous section gave you plenty of examples of the kinds of answers the examiners are looking for.

You have seen that you need a body of *knowledge* to answer these questions, but regurgitation of facts is not enough. You must also be able to show a depth of *understanding* of the facts and their application in society, and you will have to demonstrate certain *skills* and *scientific processes.*

In the final assessment, marks will be allocated in the following proportions:

Knowledge and understanding	45%
Skills and processes	55%

The proportions of marks allocated to the different themes will be:

Diversity of organisms	5% to 10%
Relationships between organisms and with the environment	25% to 40%
Organisation and maintenance of the individual	25% to 40%
Development of organisms and the continuity of life	15% to 25%

At least 15% of the marks in the complete examination will be allocated to the application of biology in modern society.

The scheme of assessment will be divided into two parts: a written examination and an internal assessment of experimental skills.

The written examination accounts for a maximum of 80% of the final mark. It consists of one or two papers.

With some groups, one paper is taken by all candidates, and there is a second, extended level paper which is optional. This enables a candidate to raise the grade achieved on the first paper by one or two grades. Candidates taking the first paper can gain a maximum grade C, but grades A and B can only be achieved by candidates taking the extended level paper. This is the scheme operated by the SEG.

Other groups, e.g. LEAG and the NEA, have one paper to be taken by all candidates. They then opt for a more straightforward second paper with a maximum grade C, or an extended level paper with a maximum grade A.

If you compare the questions on the papers with a maximum grade A with those with a maximum grade C, you will see that those at the extended level require a greater depth of understanding and more skills in experimental work.

Turn back to section 4 and compare the genetics questions set by the NEA. In question 12 on Paper IIP (maximum grade C), on page 36, the candidate has to show an understanding of the idea of dominance and apply it. In question 13 on the extended level paper, IIQ, on page 40, the candidate must understand both dominance and codominance; practical techniques in carrying out an experiment must be described and the results presented and analysed.

The questions on the maximum grade C paper set by the LEAG (Paper 2) are more structured than at the extended level. To see this, compare the question on the skin, ***3A*** in section 5 on page 69, with the following question at the extended level:

Explain how the skin helps to return body temperature to normal after vigorous exercise. *(14 marks)*

For this question, the candidate has to select and organise relevant data into a logical sequence to give a complete answer. All aspects of the question must be covered, as the total marks will be divided between the different areas. In this example, both sweating and increased blood flow to the surface of the skin must be explained. If you described sweating only, you would not be able to gain full marks, however detailed your answer.

Generally each relevant point you make will score one or half a mark up to the maximum for that question. You will not be given credit for any additional points you make, so do not spend too much time on any one question at the expense of others.

The written examination tests knowledge, understanding and certain skills and processes. Experimental skills and some observational work can only be assessed in the laboratory.

The internal assessment of experimental skills accounts for a minimum of 20% of the total mark. The skills assessed fall into a number of domains, as shown in the table below:

Domain	**Skill**
Measurement	Reading scales
	Measuring out volume
	Scaling, i.e. working out size from a scale
	Systematic counting

Observation	Matching or using keys Describing Finding differences
Handling apparatus and material	Assembly of apparatus Handling of apparatus Carrying out safe working procedures Following written and diagrammatic instructions Using the compound microscope Preparing and staining mounts
Recording	Using diagrams to show experimental set-up or specimen Completing prepared tables Devising an appropriate table Plotting a graph on given axes which include the scale Constructing a graph
Data and its interpretation	Extracting information from tables, charts or graphs Identifying anomalous results Identifying experimental error and explaining variation in results Calculations involving addition, subtraction, averages, percentages, ratios Formulating generalisations consistent with data
Design	Criticising experimental design Formulating simple hypotheses Devising an experiment to test a hypothesis

You will acquire these skills during the practical work of the course. When you have had sufficient practice, you will be given an exercise to demonstrate each particular skill.

If you do this satisfactorily, no further assessment of that skill will be necessary. Your teacher will record your achievement on your record sheet. If you do not show proficiency at that stage, you may be reassessed later in the course.

All candidates should be able to master some of the skills during the course, but the degree of proficiency in each domain will vary.

The teacher will send the completed record card of each candidate to the examining group at the end of the course. The various groups have different methods of moderating the marks submitted by the teachers. They will allocate a final mark to each candidate for his or her practical skills.

Each candidate's final grade will be awarded on the marks for the written papers and the internal assessment of practical skills.

GRADE DESCRIPTIONS

Grade descriptions are provided to give a general idea of the standard of achievement of a candidate awarded a particular grade.

The grade awarded will depend on the extent to which the candidate has met the assessment objectives overall. A weakness in one aspect of the examination may be balanced by an above average performance in some other.

Candidates awarded a grade F would satisfy most objectives at a simple level, demonstrating basic knowledge and understanding and some of the skills of biology. They would have sufficient knowledge to understand a straightforward biological statement or problem and to be able to follow routine procedural instructions.

Candidates achieving grade C would have a deeper knowledge, to allow insight into the significance of a biological statement or problem. They would be able to plan experiments and draw conclusions from their results.

A more detailed description of grade F and grade C is given in the National Criteria as follows:

Grade F	***Grade C***
Candidates are likely to be able to demonstrate the ability to:	Candidates are likely to be able to demonstrate the ability to:
1 recall basic biological facts and principles;	*1* recall a wide range of biological facts and principles;
2 describe a limited range of simple biological experiments;	*2* describe a wide range of biological experiments;
3 understand simple safety precautions in laboratory work;	*3* understand simple safety precautions and explain the principles underlying them;

4	recall simple applications of biology in modern society;	*4*	recall simple applications of biology in modern society and explain the principles underlying them;
5	make simple observations from a variety of sources (e.g. drawings, photographs, graphs);	*5*	make accurate and detailed observations from a variety of sources (e.g. photographs, drawings, graphs);
6	draw simple inferences from biological information;	*6*	analyse and interpret biological information;
7	use a knowledge of biological processes and principles in familiar situations;	*7*	use a knowledge of biological processes and principles in familiar situations and apply it to unfamiliar situations;
8	suggest simple explanations for biological phenomena;	*8*	make hypotheses in testable form;
9	suggest and conduct simple experiments to test hypotheses;	*9*	plan and conduct experiments to test and refine hypotheses;
10	criticise the design of simple experiments;	*10*	criticise the design of simple experiments and suggest improvements;
11	use sufficient biological terms to avoid misunderstanding when communicating information;	*11*	use essential biological terms accurately when communicating information;
12	present biological information and ideas in a straightforward manner.	*12*	present biological information cogently in a variety of ways.

Items 1–4 and 11 relate to knowledge and understanding. Items 4–12 relate to skills and scientific processes.

The grades range from grade G, where the candidate has not satisfied most objectives, to F; E; D; C; B and A. Candidates awarded a grade A would show the highest degree of understanding and skills which can be expected at this level.

SUMMARY

After a gestation period of nearly twenty years, the GCSE has been born. To sum up, it is a course which:

1 forms a basis for further studies in biology;
2 is relevant to the problems of industry and society;
3 has a practical approach, so that skills and processes are learnt, as well as knowledge and understanding;
4 consists of a single system of examinations, so no early decisions have to be taken as to which course you follow;
5 meets the demands of the National Criteria.

Perhaps the most encouraging aspect is that the examination is designed to give you positive credit for your achievements rather than negative marks for your weaknesses.

So enjoy your course and good luck in your exam.